THE RIVERS
DODDER & PODDLE

MILLS, STORMS, DROUGHTS AND
THE PUBLIC WATER SUPPLY

DON MCENTEE & MICHAEL CORCORAN

Published by Dublin City Council
Civic Offices
Wood Quay
Dublin D08 RF3F
Ireland
www.dublincity.ie

Third title in the Dublin City History Engineering Series
Series editors Mary Clark and Michael Phillips

Distributed by Four Courts Press
7 Malpas Street
Dublin D08 YD81
Ireland
www.fourcourtspress.ie

Hardback edition   ISBN 978-1-907002-24-3
Paperback edition   ISBN 978-1-907002-27-4

Designed by VERMILLON
No. 1 The Stables
Distillery Lofts Design Studios
Distillery Road
Dublin 3
Ireland
www.vermilliondesign.com

Printed by Printrun Ltd
Unit 72
Western Parkway Business Park
Ballymount
Dublin 12

FOREWORD

The presence of water has always attracted human activity to its proximity, particularly if it is a river. Water flowing as a river can provide great benefits to dwellers and industry within its catchment but it can also be a source of tragedy when it is flowing in torrents or when it dries up and no water is available. A clean water supply is critical to the survival in any community and a continuous supply is essential to maintain local economy through the use of it to operate mills and for use in irrigation.

Dublin has a large number of rivers, many of which are underground. The three largest rivers are the Liffey, Dodder and Tolka, all of which are open throughout their length. Each river has its own attributes and peculiarities but all of them have served the city well or in earlier times, the Townships which are now incorporated into the city. This book about the Dodder tells a very special story about a river which exhibits all the goodness and heartbreak of a child. It can be a gentle flowing river, full of fish, and enhancing many parks on its way to the sea or it can turn into a roaring torrent, flooding houses and ripping up trees even though it has two dams to control it close to the source. Throughout its history it has provided water to the mills, monasteries and farmers. In addition it provided a water supply to the Township of Rathmines and Rathgar which finally became part Dublin City in 1930. This of course created competing interests for the limited available water and it is interesting to read the outcomes.

Thus the River Dodder has played a critical part in the growth of Dublin and has adapted to the varying needs of the population down through the ages. We are very fortunate to have as two authors, Don McEntee and Michael Corcoran, who have spent a large part of their working life with the City Council and know the importance and relevance of such a river. Their association with the Council makes this book a personal journey and I am extremely grateful that they are sharing this story with us. While the story of the Dodder, like any other river, will continue to impact on the lives of people into the future this publication helps us to understand and appreciate the importance of the river in people's lives who have gone before us and assist us in appreciating all the more the value of the river for future generations.

MICHAEL PHILLIPS
Dublin City Engineer

ACKNOWLEDGEMENTS

Assembling the material for this work necessitated extensive research – and much pleasure derived from the various books, papers, maps, drawings and other documents studied in the process. Foremost among the sources consulted were C.L. Sweeney's *The Rivers of Dublin* and Christopher Moriarty's *Down the Dodder*. Also very informative was *Dublin City and County – From Prehistory to Present* (Aalen and Whelan, editors). William Nolan's *Society and Settlement in the valley of Glenasmole 1750-1900* (Chapter Seven) supplied invaluable historical information. Michael Murphy's comprehensive paper and assistance covering the Rathmines and Rathgar water supply proved indispensable, as did Seamus O Maitiu's *Dublin's Suburban Towns* and various papers and publications by the late Patrick Healy. Several other publications studied or quoted are acknowledged in relevant sections and contexts throughout these pages. John Lennon has provided valuable historical assistance and compiled the section on milling on the River Slang. Angela O'Donoghue provided useful information on the mills in the Owendoher catchment. The City Council's Library service and the Met Eireann Librarian supplied a wealth of weather records. Tom Curran and Michael Victory of Surveying & Mapping, Dublin City Council provided valuable assistance in locating and preparing suitable maps for inclusion in this book from their collection. Sile Coleman of South Dublin Library made available maps from their extensive collection. Nick Ryan provided numerous publications on milling. He proof read the first rough draft of this book and his extensive comments were invaluable in finalising the book. Hugh Fogarty researched the origin of some of the Gaelic place names and rivers and the Ancient Irish Laws relating to water rights and mills. Brian Siggins reviewed the section on Ringsend. Michael Phillips researched the construction of the locks at Grand Canal Basin and Ringsend Bridge. Gerry O'Connell reviewed and updated the section on flood risk and flood alleviation works on the Lower Dodder. Ger Goodwin and his staff in Bohernabreena did a great cleanup job on of the old waterworks and provided some excellent photographs of the river in flood conditions. Colm McEntee provided expert assistance in restoring a number of the old maps.

Historical sources are listed in the Bibliography at the end of this book, and it should be pointed out that some of these yielded differing historical and statistical information even when published around the same time.

CONTENTS

GLOSSARY OF TERMS

AIR VALVE is an artificial (man-made) channel that is constructed to convey water from one location to another. This can be a canal, pipeline, inverted siphon, reservoir and tunnel.

AQUEDUCT is an artificial (man-made) channel that is constructed to convey water from one location to another. This can be made of canals, pipelines, inverted siphons, reservoirs and tunnels.

ASTRO TIDES are forecast tide levels influenced solely by heavenly bodies, while real tide levels are dictated by many other factors, particularly atmospheric pressure and winds.

EDUCTION TUNNEL is a tunnel formed underneath a dam. Initally the river is diverted through the tunnel to permit the construction of the dam embankment. The delivery pipes from a reservoir are laid through the tunnel. Before the reservoir is filled the tunnel is sealed.

FOREBAY is an artificial pool of water in front of a larger body of water. The term in the book refers to the water surrounding the takeoff tower. This is the deepest part of the reservoir.

FIRE PLUG The concept of fire plugs dates to at least the 17th century. This was a time when firefighters responding to a call would dig down to the wooden water mains and hastily bore a hole to secure water to fight fires. The water would fill the hole creating a temporary well, and be transported from the well to the fire by bucket brigades or, later, by hand-pumped fire engines. The holes were then plugged with stoppers, normally redwood, which over time came to be known as fire plugs. The location of the plug would often be recorded or marked so that it could be reused in future fires. This is the source of the colloquial term fire plug still used for fire hydrants today.

MILL POND is a body of impounded water used as a reservoir for a water-powered mill. The water level is higher than the mill

MILL RACE An artificial water channel constructed beside a river to conduct water from the river to a mill pond or directly to a waterwheel.

MULCHURE A toll to which millers are entitled, (the 24th part of the grain ground) in lieu of a money payment.

PRESSURE HEAD In fluid dynamics *head* is a concept that relates the energy in an incompressible fluid, (e.g. water) to the height of an equivalent static column of that fluid. For a waterwheel the energy from the water level in a mill pond relative to the bottom of the waterwheel is the pressure head available to turn the waterwheel and convert this energy into motive power. Head is expressed in units of height such as metres or feet. In a waterwheel the height of the wheel combined with the width of the wheel determines the power output from the waterwheel.

SADDLE BLOCKS were used when boards were cut by hand from a round log with a pit saw. To achieve this the log was put on wooden blocks, similar to trestles, to hold it off the ground. These blocks were call saddle blocks.

SCOUR VALVE on a water main is usually located at a low point on the main. It is used to drain down the water in the main in an emergency or for maintenance. If there is a build up of deposits in the water main this is removed through the scour valve.

SHIP MILL/BOAT MILL is a type of watermill. The milling and grinding technology and the drive (waterwheel) are built on a floating platform on this type of mill. Its first recorded use dates back to mid-6th century AD in Italy

SLUICE/FLUME is wooden trough controlled at its head by a gate channelling water toward a waterwheel.

SLUICE GATE is traditionally a wood or metal barrier sliding in grooves that are set in the sides of the waterway. Sluice gates commonly control water levels and flow rates in rivers and watermills. The term "sluice gate" refers to a movable gate allowing water to flow under it. When a sluice is lowered, water may spill over the top, in which case the gate operates as a weir. Usually a mechanism drives the sluice up or down. This may be a simple hand-operated worm drive or it may be electrically or hydraulically powered.

SPILLWAY A spillway is a structure used to provide the controlled release of flows from a dam. The *spillway crest* is the top of the spillway.

STILLING BASIN at the terminus of a spillway serves to further dissipate energy and prevent erosion. It is usually filled with a relatively shallow depth of water and sometimes lined with concrete. A number of velocity-reducing components can be incorporated into the design to include chute blocks, baffle blocks, wing walls or an end sill.

TAIL RACE is a channel carrying water that has passed over or under the waterwheel back into the river or to another mill race feeding another mill.

TUCK MILL was used in the woollen industry for cleansing of cloth (particularly wool) to eliminate oils, dirt, and other impurities, and making it thicker.

TUMBLING BAY is the collection basin situated below a weir and at the head of a spillway.

VALVE is a device that regulates, directs or controls the flow of water by opening, closing, or partially obstructing various passageways through which water flows.

WEIR is a barrier across a river designed to alter its flow characteristics. In most cases, weirs take the form of obstructions smaller than most conventional dams which cause water to pool behind them, while allowing water to flow steadily over their tops. Weirs are commonly used to alter the flow of rivers to prevent flooding and measure discharge. When used with a watermill a weir is used to direct water into a mill race built upstream of the water mill.

WHIPSAW/PITSAW was originally a type of saw used in a saw pit, and consisted of a narrow blade held rigid by a frame and called a frame saw. It was used to reduce large logs into beams and planks. Sawyers either dug a large pit or constructed a sturdy platform, enabling a two-man crew to saw, one positioned below the log called the pit-man, the other standing on top called the top-man. The saw blade teeth were angled and sharpened as a rip saw so as to only cut on the downward stroke. This arrangement made it easier for the man above to raise the saw, thereby reducing fatigue and backache - the sawyers worked together to raise, lower, and guide the saw.

INTRODUCTION

ver many years the River Dodder, rich in history and archaeology, has been the engrossing subject of numerous books and papers. Most of what has been written focuses on particular aspects of the river, e.g. flora and fauna or folklore and legend. In contrast, this book concentrates on the engineering history and topography while not neglecting other relevant issues of the river and the Bohernabreena Reservoirs. The Dodder's role in supplying water to Rathmines and Rathgar and the later integration of this system with the wider Dublin public water network is also explained.

Information has been collected from a wide range of very diverse sources – some of them contradictory - and only inserted on verification. The Bohernabreena Reservoirs, more properly known as the Glenasmole Reservoirs, and their unique role in water supply, millers' compensation rights and flood control, are a central feature of these pages. We have tried to describe the Dodder – as with any other river, having its own unique catchment and other attributes - from as many other different viewpoints as possible.

Studying or writing about any subject in isolation from its place in the wider world usually results in an incomplete and unsatisfactory overview. The benefits of adopting a broader approach will be appreciated when looking at the Dodder's tributaries, among which the Poddle is certainly the most notable. This diminutive watercourse, through which the Dodder River supplied the City of Dublin with water by proxy from 1244 until 1778, merits much more than a passing reference and is therefore a prominent component of this narrative.

The Dodder's long recorded history and the diversity of the roles it has filled - and including the havoc it has caused over the centuries – richly merits exploration within the various approaches adopted in this book. The compilers hope that the Dodder's many characteristics examined in the following pages will provide an acceptable record and stimulate others to undertake further research.

Insofar as is possible, historical events are described chronologically, but it has sometimes been necessary to conclude a particular topic before going back in time to introduce another. Fords and bridges, mills and millraces, weirs, villages and other notable features (extinct or extant), with direct relevance to the river, are described.

Although little is known of the remote history of the Dodder, some sadly incomplete records survive of mills that worked in the thirteenth century. Considerably more is known about the industrial development of the river and its tributaries that began in the late seventeenth century. Until the late 1800s water, where available, was the preferred power source for most mills and factories. Apart from the mills, the only development near the river during this period was the erection of some cottages by employers for their workers. The only other happenings of note were the floods which occurred from time to time.

Upstream from Ringsend, the only villages or hamlets of note were Ballsbridge, Donnybrook, Milltown, Rathfarnham, Templeogue and Tallaght, During the nineteenth century autonomous townships were established outside the Dublin City boundary, at that time defined mainly by the canals. On the southside, Rathmines & Rathgar and Pembroke, grew rapidly, approaching the Dodder from 1850s. Urbanisation eventually reached and later embraced the river.

West of Rathfarnham, the Dodder's surroundings remained stubbornly rural, with little or no development around Templeogue or Tallaght. It was hoped that the Dublin and Blessington Steam Tramway, opened in 1887, would encourage building, but when the line closed in 1932, nearly twenty years would elapse before this came to pass. Today, the Dodder, a much loved amenity, flows through attractive linear parks in built-up areas, kept pristine by volunteers who carry out regular cleansing sessions.

Some confusion can arise from the different idiom and nomenclature used by people writing in long separated eras. In several instances, house names were changed or the premises referred to were demolished or replaced since the earlier articles or books were written. Place names have also been altered and two or more titles have sometimes been used simultaneously for the same area or district. To locate and understand fully the places described by people who wrote at different times, contemporary maps are extremely informative. It is also worth bearing in mind that perceptions and attitudes - official, communal and individual - have changed from time to time. So, what happened in any particular period should, as far as possible, be considered in the light of the mores and circumstances of that era.

CHAPTER 1

MAPS – SCALES, LEVELS AND BENCHMARKS

aps and drawings are indispensable tools in the study of rivers. Comprehensive collections of these documents, showing the natural and man-made physical features in their areas, are an essential component of local authority records. In the offices of Dublin City Council and South Dublin County Council, Ordnance Survey maps showing details of the Dodder, its tributaries and environs, are updated regularly. A brief explanation of the Ordnance Sheets, and their scales, levels and benchmarks may be helpful in understanding much of what follows.

The earliest published Ordnance Survey maps in the 1830s were drawn to a scale of 1:10560 or six inches to the mile. Town maps to ten times that scale (1:1056) were also produced. The City Council's Drainage Division has a complete set of the 1864 edition, with 33 sheets showing the City within the municipal boundary of that time.

The Drainage Division also has the 1887 and 1908 editions; more extensive in the area they cover and gridded differently from their predecessors. These three sets of 1:1056 maps, (60 inches to the mile or 88 feet to the inch) are collectively known as Eighty-eights. For those who have worked with maps over the years, these Eighty-eights are regarded as an invaluably and consistently accurate record of the nineteenth and early twentieth century city and its suburbs.

In time, maps laid down to the scales of 1:2500 (approx. twenty-five inches to the mile) and 1:1250 (approx. 12.5 inches to the mile) were also issued. While most of these are now obsolete, they are preserved as a record of historical topography. All these old ordnance sheets, but particularly the eighty-eights, are still a primary source of reference.

All the maps just described display Imperial measurements with levels and benchmarks based on Poolbeg Datum (PbD). Until the late 1960s, levels throughout Ireland were related to this datum. "The Altitudes are given in Feet above the Low Water of Spring Tides in Dublin Bay, which is 21 feet (6.4 metres) below a mark on the base of Poolbeg Lighthouse" was part of the information printed on old Ordnance maps. A most important development in the late 1960s was the changing of the Datum location from Poolbeg to Malin Head, the difference in levels between the two places being 2.714 metres.

In 1966, as Ireland was preparing to adopt the metric system, the Ordnance Survey introduced their first 1:1000 scale maps. The earliest of these used a mixture of Imperial and metric measurements which could result in mistakes. So, to ensure that there would be no confusion, the City Council's early 1:1000 sheets with metric dimensions display a prominent X. Another pitfall from that era is that some Dublin maps still showed Poolbeg (PbD) datum but this was later superseded by Malin Head (MHD), The City Council's older 1:1000 maps are therefore clearly annotated to indicate the appropriate datum (MHD or PbD), alerting users to exercise special care while looking at 1:1000 maps from that time.

New Ordnance maps, architectural and engineering drawings finally concluded their long-drawn-out transition to an all-metric state in the early 1990s. But maps and drawings from the dual-system period (and earlier) still have to be read and some knowledge of the Imperial system is important for anybody consulting such documents. In many long-established architectural and engineering offices, there are drawings going back centuries which are enumerated in feet and inches, yards and perches, miles and furlongs. To interpret these accurately or convert them to metric calls for a competent grasp of the Imperial system and the metric equivalents of its key measurements. The following table sets out the scales most often found on Ordnance and other maps.

Benchmarks, indicating the relationship of particular locations to Poolbeg or Malin Head datum, were a feature of Ordnance sheets for many years. The level of the mark was printed on the map together with an arrow pointing to the exact location of a benchmark (sometimes called a crow's foot) chiselled into the masonry of a building, bridge or other permanent structure. Crow's Foot benchmarks consisted of the familiar British War Office device, an arrowhead pointing upwards against a horizontal line marking the actual level. From the 1960s onwards, the crow's foot benchmark gave way to the brass stud and, later, to small brackets driven into walls. While all of these items now belong to the past - swept away by new technology - a diminishing number of crow's foot bench marks survive, some of them on bridges over the Dodder.

Scales Used On Ordnance And Other Maps

Imperial Scale	Metric	Notes
41.66 Feet to 1 Inch	1/500	
44 Feet to 1 Inch	1/528	
50 Feet to 1 Inch	1/600	
52.083 Feet to 1 Inch	1/625	
83.33 Feet to 1 Inch	1/1000	Standard Metric Scale
88 Feet to 1 Inch	1/1056	5' 0" (60in) = 1 Mile
104.166 Feet to 1 Inch	1/1250	
208.33 Feet to 1 Inch	1/2500	25" Maps
880 Feet to 1 Inch	1/10560	6 Inches to 1 Mile

In these pages, measurements and statistics quoted from before 1974, when the S.I. metric system was officially adopted, are Imperial, followed by the metric equivalents in brackets. In the S.I. system of linear measurement, only millimetres and metres were to be used and the centimetre ignored.

The 6-inch maps (in black & white or colour) of the 1830s can be accessed on the Ordnance Survey website at http://www.osi.ie/ or put OSI in Google search. Open OSi.ie -National Mapping and click on MapViewer. Click on GeoHive and find the location you are interested in by continuous clicking on the map until enlarged map appears. On the menu Click on Base Information and Mapping. Then click the HISTORIC 6 inch colour, HISTORIC 6 inch B&W (surveyed in 1837 and published in 1843) or HISTORIC Map 25 inch (published in 1912) and view the location as it was in the nineteenth century. A facility exists in which one map can mesh into the other using "Transparancy" slider at the bottom. In this way, a researcher can identify a particular modern day location and incrementally see how the area looked in the 1830s in the six inch maps and the early twentieth century in the twenty five inch maps. The latest ORTHO aerial photographs can also be viewed here.

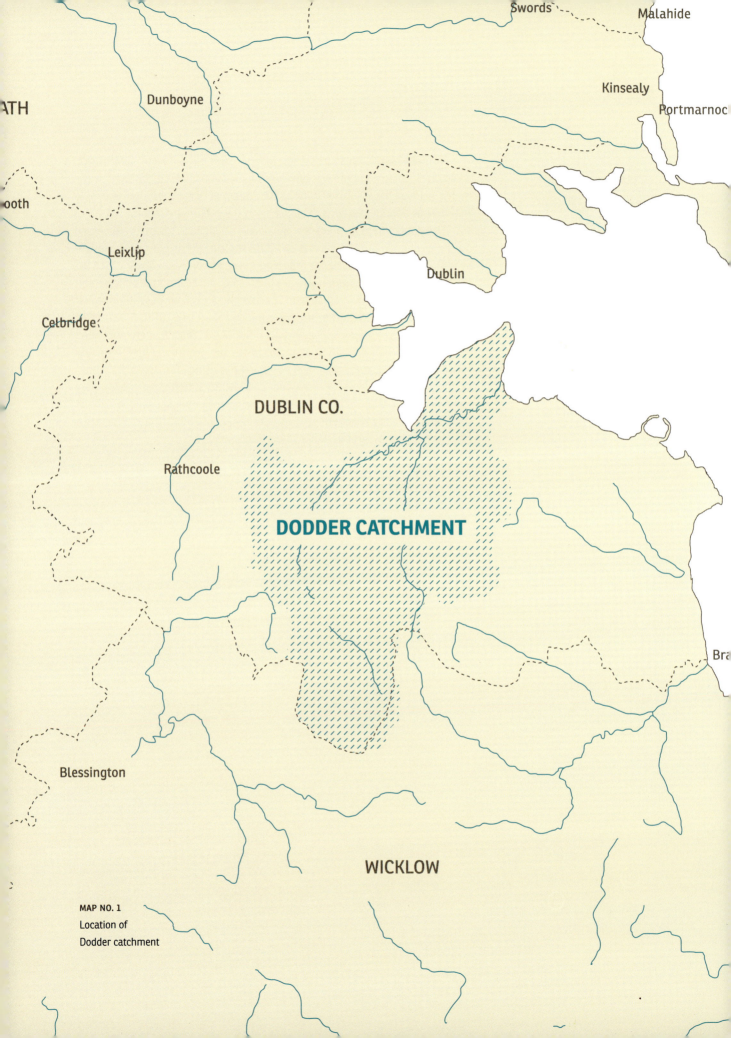

ATH

Swords
Malahide

Dunboyne
Kinsealy
Portmarnoc

ooth

Leixlip
Dublin

Celbridge

DUBLIN CO.

Rathcoole
DODDER CATCHMENT

Bra

Blessington

WICKLOW

MAP NO. 1
Location of
Dodder catchment

CHAPTER 2

GEOGRAPHY OF DODDER VALLEY

Graystones

Kilcoole

Newtown
Mount
Kennedy

Throughout time, the Dodder [MAP NO. 1] has had several names, some of them purely local. It has been known variously as the Dodere, Dodeer or *Dothair* and, in Prince John's Charter of 1192, the Dother. The names given are all variant versions of the Irish name *(An Dothar)*. The earliest mention of the name known to Hugh Fogarty is in the early Irish saga *Togail Bruidne Da Derga* ('The Destruction of Da Derga's Hostel'). Segments of the saga are found in Leabhar no hUidhre, the earliest manuscript containing secular saga-texts, which can be dated to the earliest years of the twelfth century AD (but the language of the texts is older than 1100—sometimes considerably so).

There are also mentions of the river in glosses to *Feilire Oenguso* ('The Martyrology of Oengus'). The Old Irish text was composed in the ninth century, but the glosses are probably at least a couple of centuries later. There are also references to the river—by the name *Dothra*—in the *Metrical Dindsenchas* (a collection of Middle Irish poems about the names of famous or prominent places), and in the *Martyrology of Gorman, Lebor na Cert* (the 'Book of Rights'), and the Annals of Ulster

In a number of instances, the names of places along the Dodder have been given to the river at various times, notably Rathfarnham and Donnybrook; various spelling differences will also be encountered. An American literary researcher who visited the Drainage Division in the 1970s was convinced that Dodder was an onomatopoeic title bestowed on the river by James Joyce to describe the meagre and somewhat erratic flow along its stony bed during prolonged dry spells.

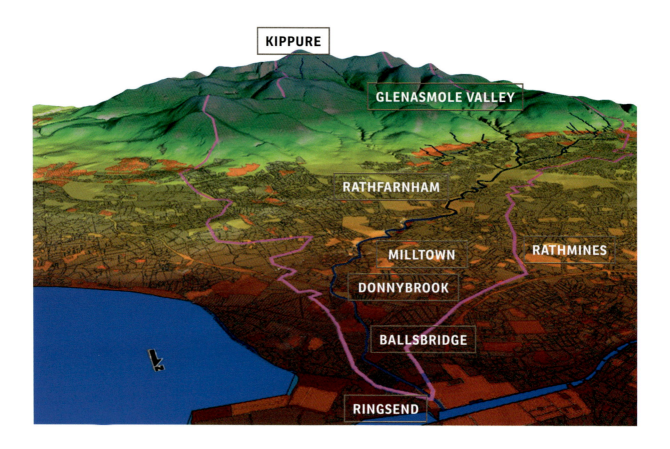

KIPPURE

GLENASMOLE VALLEY

RATHFARNHAM

MILLTOWN

RATHMINES

DONNYBROOK

BALLSBRIDGE

RINGSEND

The Dodder [AS SHOWN IN FIG. 1 – MONTAGE OF THE DODDER CATCHMENT] has its beginnings in the chain of hills bounding on the south of County Dublin, separating it from Wicklow and known as the Dublin Mountains. At about a mile and a quarter (2km) to the north of the peak of Kippure, the sources of the Dodder are only four miles (6.4km) from those of the Liffey. However, while the Dodder takes a more direct course of less than 17 miles (26.9km), the Liffey follows a longer, circuitous path of about 85 miles or 137 km through three counties and Dublin City before the two rivers meet at Ringsend. It is noteworthy that, according to the measurement protocols used, divergent lengths are quoted for both rivers in various books and reports.

Some of the Dodder's waters come from land lying near the summits of mountains in Wicklow. West of Lough Bray, there rises Tromanallison, which soon joins the Inchoate River, next augmented by Mareen's Brook, the largest of the Dodder's principal constituents. It rises behind Lough Bray on Kippure, the summit of which is 2,473 feet (753.8 metres) over Poolbeg Ordnance datum. Mareen's Brook then courses down from Glassamucky Brakes, passing Heathfield, to Castlekelly, where it is joined by the Cot Brook from Castlekelly Bog (Barnachiel) and the Slade from Glassavullaun. Rising near the summits of Kippure, Barnachiel and Glassavullaun mountains, these streams are watered by the vast area of deep bog lying on their flanks. Their

FIG. 1

Montage of River Dodder catchment

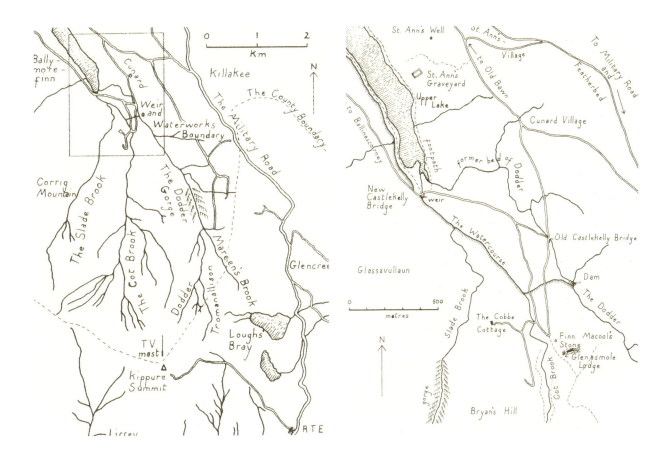

MAP NO. 2

Map of rivers in
upper Dodder

MAP NO. 3

Details of Dodder and
tributaries at upper
reservoir, Bohernabreena

courses and other features are described in detail by Christopher Moriarty [MAPS NOS. 2 & 3]. In the past, the upper part of the Dodder was known as The Cataract of the Rowan Tree.

In 1844 the illustrious engineer Robert Mallet (1810-1881) wrote: "Here, properly commences the river Dodder, which, passing, with a rapid fall, through the whole extent of this valley, hemmed in on either side by lofty and precipitous hills, covered with boulders, alluvial gravel, and clay nearly to their summits, and whose watershed is peculiarly rapid and destructive, it debouches on the basin of Dublin." Mallet, who took a keen interest in the river, studied it and knew it intimately, features in later part of this book. Diversions and other man-made features of the Dodder that took place after Mallet's time will also be described in the section on the Bohernabreena (Glenasmole) reservoirs.

The frequently turbulent Dodder — described by a writer in the *Templeogue Telegraph* as ferociously precipitous — flows through the lonely and isolated valley of Glenasmole. Only about nine miles from O'Connell Bridge, this remote area was probably the last primordial Gaeltacht in County Dublin where Irish was still spoken in the 1800s and understood well into the twentieth century. The area is also rich in mythology describing the feats and travails of Finn McCool and Oisin.

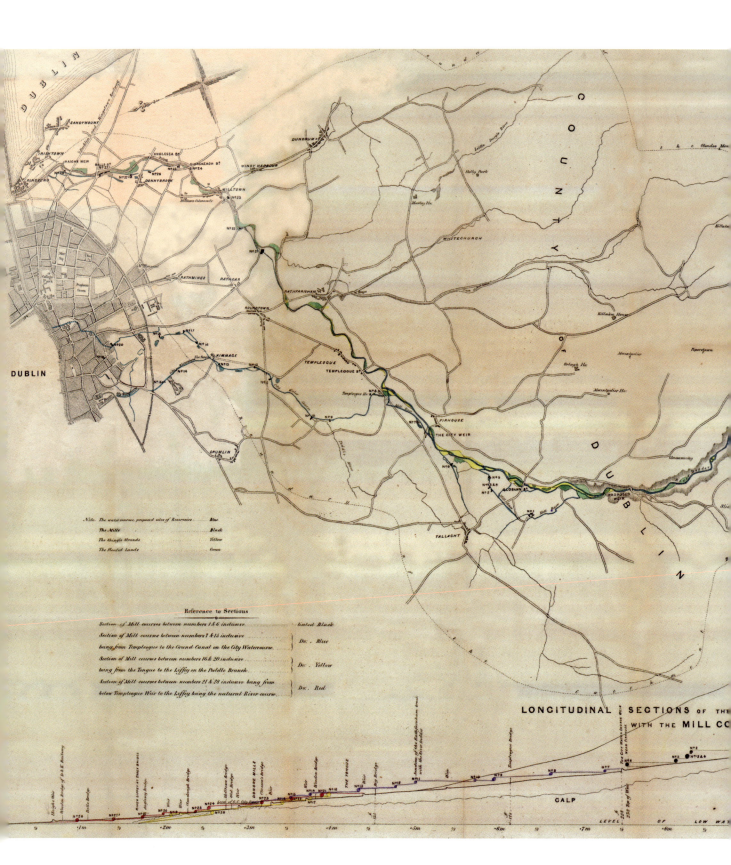

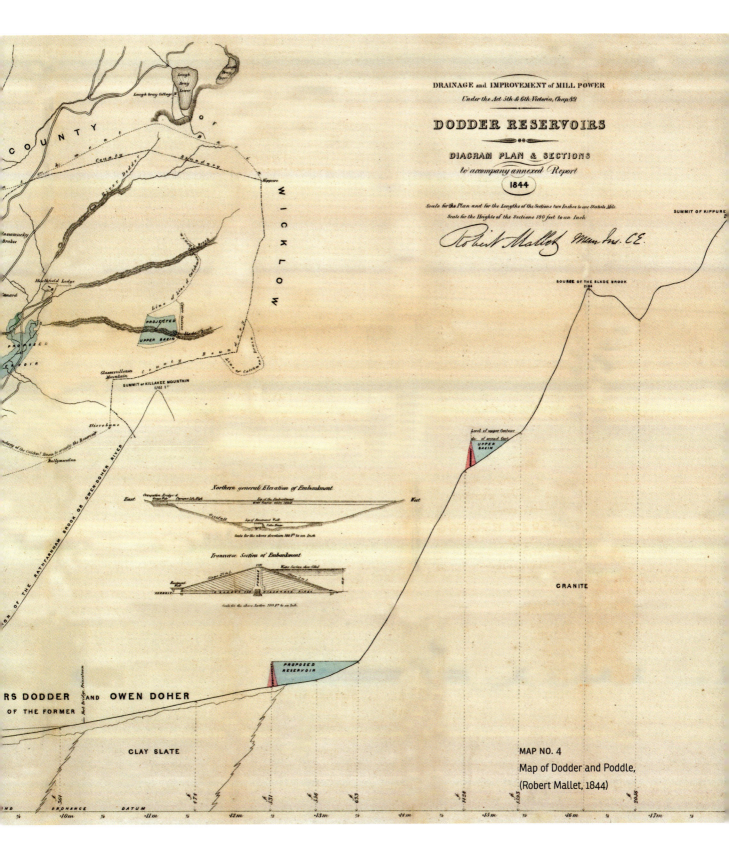

MAP NO. 4

Map of Dodder and Poddle,
(Robert Mallet, 1844)

In the early Christian era, monks were drawn to Glanasmole by its solitude and tranquillity. The reclusive St. Santan, said to have been the son of a British king, founded a monastery here in the sixth century. It appears to have survived the Viking era and is mentioned in the Annals of the Four Masters. Despite many setbacks, its church is believed to have been in use until 1547. St. Santan's name and memory disappeared for many years but, through metamorphosis, may live on in the name of St. Anne's Church in Bohernabreena. This is but one of the many stories and legends that enrich the Dodder valley.

Joined along its course by several more small streams or rivulets, shown on Mallet's MAP NO. 4 and OS MAP NO. 5, the Dodder descends through Tallaght, Rathfarnham, Milltown, Clonskeagh, Donnybrook and Ballsbridge, to discharge into the Liffey estuary at Ringsend. Its total length of 16.8 miles (26.9km) from source to outfall drains a gross catchment area of 29,902 acres (121 sq. kilometres). Older records stated the catchment to be approximately 28,000 acres (113.3 square kilometres).

Between the source of the Dodder at Kippure and the spillway crest at the Lower Reservoir at Bohernabreena (495 feet or 151m over Datum) the river falls 1,978 feet (602.9m) over 5.9 miles (9.5 km) which is a steep average gradient of 1 in 15. Downstream of this spillway crest the fall is only 495 feet (151m) in 10.3 miles (17.5km), a much gentler gradient of 1 in 115. In their report on the flood that occurred on 1st February 2002, Gerard O'Connell and Victor Coe noted that flood water takes about two hours to get from Bohernabreena to the City and that the Dodder normally has a two to three hour flash peak. Following a very heavy rainfall in the upper catchment below Kippure the level of storm water in the Dodder rises very quickly and the river reaches its peak flow within two to three hours. This is the reasons that the Dodder is known as a flashy river.

In the late nineteenth century, William Handcock estimated the catchment of the Dodder as 55 square miles (141.1 sq. kilometres), of which about 22½ sq. miles (57.6 sq. km) were mountainous and 32½ sq. miles (83.5 sq. km) plain or of moderate inclination. He referred to the many attempts by riparian owners to reclaim parts of the extensive strands—natural flood plains—between Kiltipper and Firhouse. Costly defences were erected but, in time, were inexorably undermined and levelled by the river and the tracts of reclaimed ground were devastated.

Patrick Healy records that, late in the eighteenth century, Ponsonby Shaw of Friarstown (formerly Friarsland) incurred great expense to improve his property. He constructed walks, grottoes and waterfalls and also built a dam about forty feet (12 metres) high across the glen to create an ornamental lake. The dam burst shortly after its construction and destroyed the work that had been done. A later owner was Captain Bayley, who carried out repairs, but the lake subsequently silted up. It finally became a refuse dump during the 1970s and closed in 1997.

On the lower ten miles (16km) of the river there are thirteen weirs of varying heights, which tend to curtail in some measure the rapid channel velocities that might be expected from the Dodder's average gradient. The present day weirs on the Dodder are located upstream of former watermills. In some instances

MAP NO. 5
River Dodder with main tributaries and principle towns and villages

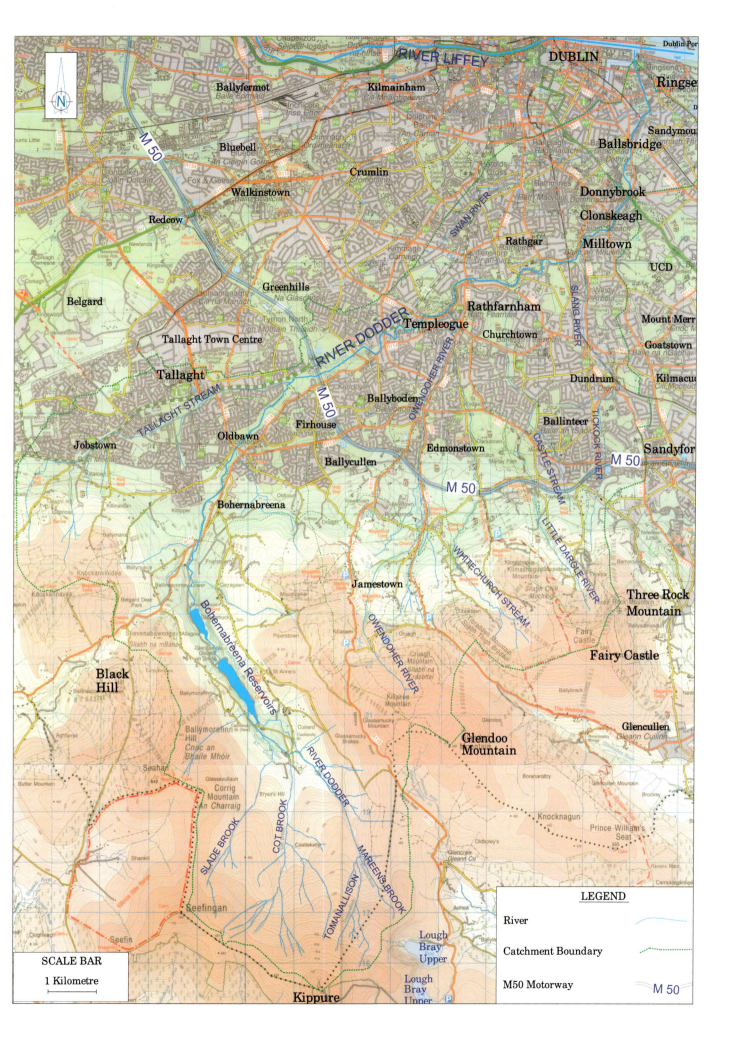

the remains of the mill races can be seen on one or other side of the river. Some of these weirs are in a bad state of repair with no one body responsible for their upkeep. Originally the mill owners looked after the weirs to ensure a supply of water to the mills. The Local Authorities have no direct responsibility in looking after them with the exception the Balrothery Weir which is maintained by South Dublin County Council and Orwell Weir which is maintained by Dublin City Council. A gauging station, to measure the flow in the Dodder, is maintained by Dublin City Council upstream of Orwell Weir.

As will be deduced from the many developments to be described later, the physical state and surroundings of the Dodder in the early twentieth century are very different from those of the late nineteenth and early twentieth centuries. The abandonment and disappearance of several millraces and the more controlled flow of water resulting from the construction of the Bohernabreena Reservoirs are two factors.

Others include the deepening of the channel above Firhouse by the removal, during the late nineteenth century, of much stone and shingle, described by Handcock as "an inexhaustible supply of rather bad road material" and which he estimated at some 4,000 tons per annum. This practice had also been condemned by Robert Mallet, who identified it as a factor in the undermining of weirs and retaining walls.

Another major element in changes affecting the Dodder and other rivers was the evolution of industry. Riverside locations became less attractive as industries grew less dependent on water power.

CHAPTER 3

PUBLIC BODIES –
DUBLIN LOCAL AUTHORITIES –
BOUNDARIES

The Dodder flows through or adjacent to the areas of three local authorities. An outline of their history, how they evolved and how they operate provides much useful background to any study of the river.

From 1840 until January 1901, the Royal and Grand canals, together with the North and South Circular roads, constituted much of the Dublin city boundary. Outside this cordon, autonomous townships were established by statute from 1834 onwards. Residents suitably qualified by property valuations and other criteria were eligible for selection as commissioners, and were responsible for the management and improvement of the townships

Meanwhile, outside the cities, the counties were still administered by the anachronistic and deficient Grand Juries, established in the seventeenth century and typically composed of grandees, magistrates, and prominent landowners. At various periods, the townships took over many of the powers previously vested in the Grand Juries.

Much of the terrible poverty so widespread in nineteenth century Ireland originated, among other causes, with the depressions that followed the 1798 Rebellion, the Act of Union (1800) and the Battle of Waterloo (1815). There was also a constant atmosphere of agrarian unrest, and the catastrophic potato famine of the 1840s.

Appointed under the provisions of the Poor Relief Act of 1838 to alleviate a deteriorating situation, the Poor Law Commissioners divided the country into unions, made up of electoral areas and townlands, which were the smallest units in local administration. Elected Boards of Guardians administered the Unions, which tackled poverty and promoted public health with a modicum of success.

In 1872, the Poor Law Commissioners metamorphosed into the Local Government Board (ancestor of today's Department of the Environment,

Community and Local Government) and, as laid down by the 1878 Public Health (Ireland) Act, the poor law unions became rural sanitary authorities.

The Dodder flowed through the territory of the County Dublin Grand Jury and the South Dublin Union. East of Rathfarnham Bridge, along the present Lower Dodder Road, it became the southern boundary of the Rathmines Township, established in 1847 and extended three times (in 1862, 1866 and 1880) to become Rathmines & Rathgar. The Dodder also formed the border between the Pembroke Township (established in 1863) and the county as far as Donnybrook, downstream of which it was entirely within the Pembroke Township to Ringsend.

Under the provisions of the Local Government (Ireland) Act 1898, democratically elected County Councils, with enhanced powers and extensive responsibilities, took over most of the greatly diminished functions previously exercised by the Grand Juries. Rural District Councils (RDCs) and Urban District Councils (UDCs) were also created, the townships surrounding the city achieving the latter status. The Poor Law Unions survived until after the First World War of 1914-1918 (The Great War) and, following several changes, most of their functions passed to the local authorities and, later, the health boards. The Health Service Executive (HSE)and the Department of Health finally took over.

A boundary extension in January 1901 doubled the size of Dublin City, which took over large areas of the county, together with three of the former townships (Clontarf, Drumcondra and New Kilmainham). The influential Rathmines and Pembroke UDCs, however, retained their independence at that time. But in 1930, in a changed political and cultural environment, these two Urban Districts were absorbed by Dublin City in another major extension of the municipal boundary. Another momentous feature of the changes in 1930 was the appointment of Dublin's first City Manager, the distinguished and very experienced Gerald Sherlock (1876-1942).

Following this enlargement of the Corporation area, the stretch of the Dodder between Templeogue and Dartry came within the city boundary. From Dartry to Donnybrook, it marked the boundary between the city and county. The city was further expanded in 1941, 1942 and 1953, but none of these extensions affected the Dodder.

The most far-reaching reorganisation of local government in the Dublin region for nearly a century began in 1986 with significant boundary changes and important reductions in the administrative area of the city. The Dodder then became the city's southern boundary from Bushy Park to Beech Hill (Clonskeagh). The reorganisation was completed in 1994 with the replacement of Dublin County Council and Dun Laoghaire Borough Corporation by three new local authorities—the County Councils of Fingal, South Dublin and Dun Laoghaire-Rathdown. And, on 1st January 2002, Dublin Corporation was renamed Dublin City Council.

When the new councils were set up in 1994, some minor but logical modifications were made to the 1986 boundaries. Under the revised arrangements, the Dodder came within the area of South Dublin upstream of Bushy Park. Downstream of the park it now marks the border between South Dublin and the city as far as Landscape Gardens. From there to Clonskeagh Park, it is the dividing line between the city and Dun Laoghaire-Rathdown and, from Clonskeagh to Ringsend, it is wholly within the city.

In times past, local authorities operated just as inferred by their titles, the word "local" accurately inferring a lack of collaboration with their neighbours. Before the introduction of pollution control measures, a prime example of non-cooperation was the neglect of river water quality. Because there was no control, gross pollution in one council's area frequently created serious difficulties for the next authority downstream. Flooding problems could also be transferred across administrative boundaries.

From the beginning, the re-organised Dublin area local authorities (Dublin City, Fingal, South Dublin and Dun Laoghaire-Rathdown, together with Meath, Kildare and Wicklow County Councils) have co-operated closely on many programmes, including water and drainage. Subjects hitherto not dealt with and requiring the involvement of more than one local authority now arise with increasing frequency. To deal comprehensively with these situations, initially informal collaboration will, in time, lead to the establishment of more prescriptive arrangements. From 2014 the newly formed Irish Water Authotity takes over the responsibility from the Local Authorities for providing water and drainage services in Ireland.

Apart from the local authorities, several statutory bodies and various other organisations have an interest in, or responsibility for, specific aspects of the Dodder. An example is National Flood Policy, approved by the Government in September 2004, and which made the Office of Public Works the co-ordinating agency. The subject was further addressed by the European Union Floods Directive, published on 26th November 2007 and enshrined in Irish Law in March 2010.

OUTSTANDING INDIVIDUALS

It is sometimes forgotten that every achievement, event or organisation is created by people, many of whose names go unrecorded. At critical times during the nineteenth century a combination of conscientious public representatives and administrators, outstanding engineers and doctors, undertook many major projects. They lifted Ireland's reputation to new heights in an era of gloom and unrest and it is fitting that some of those associated with this story should be remembered here. The following representative selection includes such as **Thomas Drummond** (1797-1840), Sir John Gray (1816-1875); Sir Charles Cameron (1830-1921); Parke Neville (1812-1886), Spencer Harty (1838-1923) and Robert Mallet (1811-1881).

Born in Edinburgh in 1797, Thomas Drummond trained as an engineer. He worked with the Ordnance Survey in Ireland before transferring to the civil service, becoming permanent Under Secretary in Dublin Castle. Empathising with Daniel O'Connell and the struggles of ordinary people against injustice, he enunciated the principle that property has its duties as well as its rights. A workaholic, he died in 1840, the year in which the Municipal Corporations Act initiated the democratisation of local government.

Dr. (later Sir) John Gray, independent MP and City Councillor, promoted the Vartry Water scheme in the 1860s. He co-operated closely with **Parke Neville** Dublin's first full-time City Engineer, who modernised existing services and introduced new ones between 1852 and 1886. Pure piped drinking water, efficient drainage, street surfacing, lighting, cleansing and refuse collection were among the essential services that received Neville's close attention. Among the many achievements of **Spencer Harty**, Neville's successor as City Engineer, was the completion in 1906 of the main drainage system.

From the mid-1800s onwards, the causes, treatment and control of many diseases were increasingly understood. Better medicines, clean food, personal and domestic hygiene, were also essential factors in improving public health. **Sir Charles Cameron**, Dublin's Chief Medical Officer from 1886 until 1912, strove unrelentingly to improve the health of the thousands of poor people who lived in the city during that period and also frequently condemned much unfit housing.

One of Ireland's outstanding engineers of the nineteenth century who had an important connection with the Dodder was **Robert Mallet** (1810-1881). The story of his fascinating ancestry, life and achievements have been well researched and set down by Enda Leaney and Patricia M Byrne, to whom we are indebted for what follows.

Robert Mallet was born on 3rd June 1810 in Ryder's Row, Dublin, eldest child and only son among three children of John Mallet (1780-1868), plumber, hydraulic-engine maker and iron founder and his wife Thomasina (d.1861). Robert junior was a good student who showed an early interest in physical science and graduated from Trinity College in 1830. He became a full partner in his father's business in 1832 and developed it into one of the most important and successful engineering enterprises in Ireland.

Mallet senior built a large engineering works beside the Royal Canal at Cross Guns Bridge, The arrival of the railways brought huge orders and there was also an export business. The company produced a very wide range of machinery and equipment, including manual fire engines. Closed as an engineering works in 1860 when orders declined, the Phibsboro building later became the North City Mills and is now an apartment block. But the once famous firm's nameplates survive on many structures throughout the country.

After 1860, Robert devoted his genius to consultancy. His achievements, interests and awards were legion; he was one of the most famous engineers of his day and was also deeply interested in science. He studied the causes and events of earthquakes and is known as the father of seismology. When the Government wanted an eminent engineer to survey and report on the Dodder in 1843 there can be little wonder that they chose Robert Mallet.

In 1831, Mallet married Cordelia Watson, a bookseller's daughter. They had three sons and three daughters and from 1836 they lived in the famous Delville House in Glasnevin. Cordelia died in 1854 and in 1858 the family moved to Monkstown and London in 1860. Robert married Mary Daniel in 1861 and was very active until about 1874 when he became blind. He fell seriously ill in 1880, died on 5th November 1881 and is buried in Norwood cemetery.

Dr Ronald Cox of TCD has compiled a record of Mallet's papers in *Robert Mallet FRS 1810-1881* (1982).

To all of the foregoing and the many anonymous others who worked with them, we are still deeply indebted today.

CHAPTER 4

DODDER VILLAGES AND TRIBUTARIES

DODDER VILLAGES – THE 1801 STATISTICAL SURVEY

Along the course of the Dodder [MAP NO. 5] or close by there are seven former villages or hamlets, six of which have been partly or totally subsumed into the City of Dublin as suburbs. A few facets of their very different and varied histories are outlined (from west to east) in the following pages.

For anybody seeking further information, many comprehensive local histories are available, several of which concentrate on one particular village or district. The gaps that remain in our knowledge of social and industrial history are being gradually closed through diligent research by various societies and their individual members. And, as interest in our past grows and research continues, new books appear regularly.

Among the many surveys and directories containing information about mills and other factories on the Dodder is the 1801 Statistical Survey of County Dublin. It was compiled for the Dublin Society (later the Royal Dublin Society) by Lieutenant Joseph Archer with a view to improving the industrial landscape and it lists various enterprises in the Dodder villages. Many of the mills that existed in 1801 were gone by the time of the Mallet survey, carried out in 1843 and extensively quoted later in this volume, took place.

Looking back at how mills and industry generally were energised in the eighteenth century, it must be borne in mind that the major sources of power were wind and water. Steam became increasingly used from early in the nineteenth century but gas, electricity and petrol or diesel power were still far in the future. Hard physical effort, using only the most basic tools and equipment, was required of those who laboured in a very harsh environment.

TALLAGHT (Tamhlacht) first of the Dodder's former villages, was where St. Maelruan established a monastery in the eighth century. The Archbishop's residence was built there in the fourteenth century. In medieval times, Tallaght was a walled town.

Moving to more modern times, the 1801 Survey locates a corn mill, owned by a Mr. Newman, in Tallaght. Thom's Directory for 1861 gives the population of Tallaght as 312 people, who lived in 57 houses. It had a constabulary barracks and a courthouse, and local employment was provided in paper making and flour milling. In 1867, Tallaght was the scene of a Fenian uprising.

Despite being served by the Blessington Steam Tramway (1888-1932) Tallaght retained much of its village character until after World War Two. Better roads, improving transport and incipient industrialisation encouraged some residential development during the 1950s and 1960s.

Starting in 1967, Dublin Corporation built up extensive land banks in the County to accommodate future housing needs; the Tallaght area was one of the selected locations. A greatly extended water supply network and the construction of the South Dublin Drainage Scheme facilitated the large developments carried out during the 1970s and 1980s.

For a time, Tallaght was a somewhat detached suburb of Dublin. However, with the introduction of the new local government arrangements for the Dublin area, its importance was greatly enhanced. It is now a major county capital, the headquarters of South Dublin County Council being located there, as are a civic centre, hotels, several schools, a major hospital and an extensive shopping complex. The 2011 population was in excess of 60,000 people and Tallaght has enjoyed a tram service to Dublin city centre since September 2004.

TEMPLEOGUE in earlier times could best be described as a hamlet, which had a mill in the seventeenth century. It became better known in the late nineteenth century when the main depot and workshops of the Blessington Steam Tramway were established there in 1888. During the 1930s, ribbon development along the road from Terenure was the first stage in making Templeogue a suburb of Dublin. Despite this, its core, especially the main street, has succeeded in preserving the atmosphere of a village. Retaining or restoring the character of other former villages that became suburbs of Dublin has now become a welcome and widespread pursuit.

RATHFARNHAM lies to the south of the Dodder, but is still regarded by historians and others as a Dodder village. The Owendoher River, a tributary of the Dodder, which flows northwards through Rathfarnham also powered mills in this area. Four mills are listed in Rathfarnham in Archer's 1801 Statistical Survey of County Dublin. Newman, Curraghan and Flanagan owned a flour mill while Drumgold had a corn mill. There were two paper mills: one was owned by Freeman, the other by Teeling.

Thom's Directory for 1862 gave the area's extent as 2,531 acres and population as 5,555. Rathfarnham also had a courthouse and was connected to Dublin city centre from 22nd June 1879 by horse trams operating via Harold's

Cross and Camden Street. On 9th November 1899 these were replaced by electric cars which in turn gave way to buses on 1st May 1939.

Rathfarnham Castle, part of which dates from the sixteenth century, is foremost among the village's principal amenities. Rathfarnham was in County Dublin until 1953 when a boundary extension brought it under Dublin Corporation control, but the former village reverted to the County under the 1986 boundary changes. When the new local authorities took over from Dublin County Council in 1994, Rathfarnham was divided administratively from north to south. The western section including Main Street passed into South Dublin County Council's area, the eastern portion going to Dun Laoghaire-Rathdown.

MILLTOWN, situated at the confluence of the Slang (Dundrum) and Dodder rivers, is referred to as early as 1260 in an Inquisition of the Manor of St. Sepulchre. This was in connection with the murder, twenty years earlier, of a miller and the imprisonment of another miller and his sons for not apprehending the culprit. The area was subject to attack by displaced Irish clans in the fourteenth century. During the Cromwellian period the population of Milltown was given as nineteen—fourteen English and five Irish.

Quarrying flourished in Milltown in the sixteenth century, when stone was taken for the repair of Christ Church Cathedral, directed by architect Sir Peter Lewys, builder of the bridge of Athlone. With great difficulty, stone was cut out of the bed of the Dodder using iron tools which were kept pointed by a blacksmith who was in constant attendance. In order to allow the stone cutters to sever the rock, the river had to be diverted. One day a bank of earth fell on a mason, who was lucky to survive. On another occasion, the Dodder flooded to such a height that it carried away all protection for the craftsmen.

An act of appalling cruelty which took place here in November 1753. William Kallendar, who may have helped an escaped prisoner, was so severely whipped from Milltown to Dundrum that he died in Newgate Prison a few days later, mourned by his wife and five small children. The original Newgate was in the towers of the City gate that stood where Cornmarket is today. It was replaced in the mid-eighteenth century by a new prison where the park beside Green Street courthouse is today. Demolished in the nineteenth century, this was notorious for the public executions carried out on a balcony over its entrance.

In Archer's 1801 Survey, a Mr. Hunt owned two woollen mills in Milltown while three others listed are not accredited to any named owner. There was also a logwood and oil-mills, owned by Burk and Mullen. Enterprises surveyed by Robert Mallet in 1843 are more fully described in the section listing industries along the Dodder. In 1861, Thom's Directory quoted the area of the village as 185 acres with a population of 863 who occupied 132 houses. Milltown was also an important road junction affording access to Dundrum and the attractive setting of some hostelries.

In his *History of the County Dublin* (published in six volumes 1902-1920), Francis E. Ball relates that in the eighteenth century, the village, at that time the property of the Leeson family, was the location of several mills. Among them were two corn mills (at different times), a brass mill, an iron mill, a paper mill and a mill for grinding dry woods. There were also, near Milltown, a limestone quarry and a factory where "the ingenious Heavisid" made garden pots.

A boundary extension in 1880 brought Milltown into the Rathmines and Rathgar Township (Urban District from 1899) which in turn was taken over by Dublin City in September 1930. When the municipal boundary was altered in 1986 those parts of Milltown south of the Dodder came under the control of Dublin County Council, passing to Dun Laoghaire-Rathdown when the new local authorities took over in 1994.

Although the Harcourt Street-Bray railway began operating in 1854, Milltown station, located immediately north of Milltown Road, did not open until 1860. The line closed on 31st December 1958 but the nine-arch bridge over the Dodder fortunately survived. This has been carrying Luas trams, which use the track bed of the former railway, since July 2004. The Milltown tram stop is just north of the former railway station site.

Dating from 1904, the handsomely restored 100-foot (30m) chimney in the parkland immediately east of the Nine Arches houses well disguised telecommunications equipment. This chimney served the former Dublin (Dartry) Laundry which operated from 1888 until July 1982.

CLONSKEAGH (*Cluain Sceach*, the Valley of the Bushes) is adjacent to the site of the 1649 battle between the Royalists and Roundheads (Cromwellians). It was still a tiny village, north of the River Dodder, of 63 acres in 1862, when the population was 310 persons. Its isolation was mitigated by the postal service from Milltown, which gave it three deliveries per day.

Clonskeagh became more accessible with the opening of a horse tramway from Ranelagh (also called Sallymount in the nineteenth century) to Vergemount on 17th March 1879. The electric cars which on 1st December 1899 began running into the city centre—and later to Whitehall—were replaced by buses on 2nd July 1939. When the No. 11 bus service that replaced the trams was extended to Bird Avenue and still displayed Clonskea—a shortened version of Clonskeagh that originally appeared on the trams—this name was extended by popular usage to a wider area. This practice is also encountered in other suburban districts.

There are references to mills in Clonskeagh in the eighteenth century and iron mills were established nearby in the 1800s. Archer's 1801 Survey lists two iron-work wheels here, owned by Jackson and White. Three others are credited to Stokes and Company.

From 1880 to 1930, Milltown came within the jurisdiction of the Rathmines and Rathgar Township (Urban District Council from 1899). The same sequence of boundary changes that affected Milltown resulted in the portion of Clonskeagh south of the Dodder coming under the control of Dun Laoghaire-Rathdown in 1994.

DONNYBROOK (Domhnac Broc) was the site of a settlement by 1204 when the Donnybrook Fair was first held. The village appears on Rocque's 1756 map as the "Town of Donnybrook." In an issue of the Dublin Saturday Magazine

(Page 665), JR O'Flanagan MRIA gives the ancient name of Donnybrook as Dhutherbrugh, composed of the name Dhuther, meaning the Black River, and Brugh, an inn or house of entertainment. Hugh Fogarty contends that this etymology is fanciful; the name deriving from the Irish Domhnac Broc, which means either 'the church of (St) Broc' or the 'church of the badgers'

Domhnac is a common element in Irish placenames, deriving from the Latin Dominicus, -a. This means 'of the lord'. and was frequently used to mean 'Sunday' (dies dominica, 'the day of the lord'), and also 'church' (as in 'house of the lord'). Many Irish placenames contain this element (Dunshaughlin, Downpatrick etc.); it is a well-known and common component. The 'Broc' is is sometimes said to be St Broc/Broch/Brock, but there is very little support for such a figure, so it is probably either 'broc'—the Irish word for 'badger' or a prefixed use of the adjective breac, 'speckled. etc'.

Donnybrook Fair, first held in medieval times, became a very important annual event, combining both commerce and entertainment. In time, it became increasingly unruly and its demise in 1855 was largely due to the violent fighting with which the name of Donnybrook literally became synonymous.

Local employment was provided around Donnybrook by quarries and mills along the Dodder. Archer's 1801 Survey lists two cotton-wash (calico printing) mills at Donnybrook, one owned by Dillon, the other by Duffy.

Some building, mostly of large residences, took place in the eighteenth and nineteenth centuries. Donnybrook came under the jurisdiction of the new Pembroke Township in 1863, following which more orderly development took place.

From 14th March 1873 Donnybrook was connected to the city by a horse tramway which was electrified on 23rd January 1899 and extended to Phoenix Park (NCR); buses replaced the trams on 2nd June 1940. From 1929 until 1949, the site now occupied by Dublin Bus Donnybrook Garage No. 2 was the location of an extensive permanent way yard which served the entire city tramway system. There was also a quarry on the tramway property.

The large scale development that took place in Donnybrook, Ballsbridge and Ringsend in the late nineteenth century is illustrated by comparing the OS Map of 1842, MAP NO. 6, with the OS Map of 1897, MAP NO. 7 .

BALLSBRIDGE (Droichead na Dothra) The name comes from an early resident who probably built the first bridge over the Dodder at this location. This was known as Ball's Bridge, which by the twentieth century was known as Ballsbridge and gave its name to the adjoining area. Like all the other villages or settlements along the Dodder, Ballsbridge first became important as the location of mills.

From late in the nineteenth century, more advanced industrial development took place on the west bank of the Dodder below Ballsbridge; The village had an area of 33 acres and a population of 573 in 1862. In the following year, it fell within the area of the newly established Pembroke Township, which became an Urban District in 1899 and was taken into Dublin City in 1930.

Most industrial development in the Ballsbridge area is described in the section on industries along the Dodder. But, on the west side of Shelbourne Road,

MAP NO. 6 (OVERLEAF LEFT)
Map of Rindsend
& Donnybrook (1837)

MAP NO. 7 (OVERLEAF RIGHT)
Map of Ringsend & Dodder
River 1897 illustrating the
extensive development
in the area since 1837

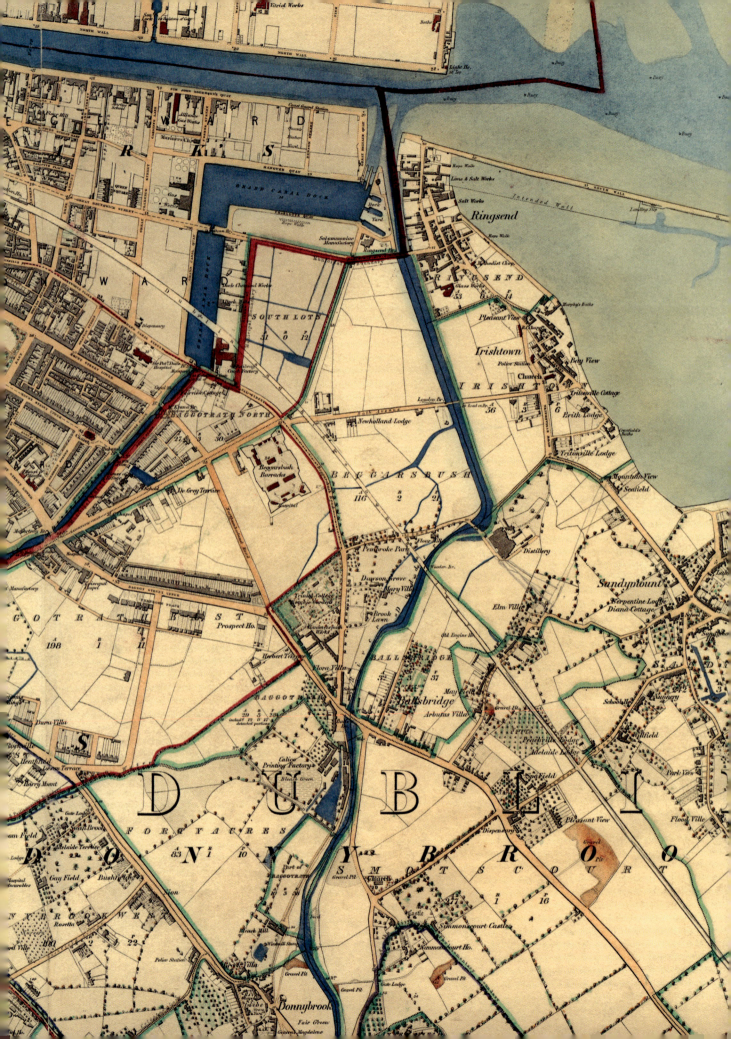

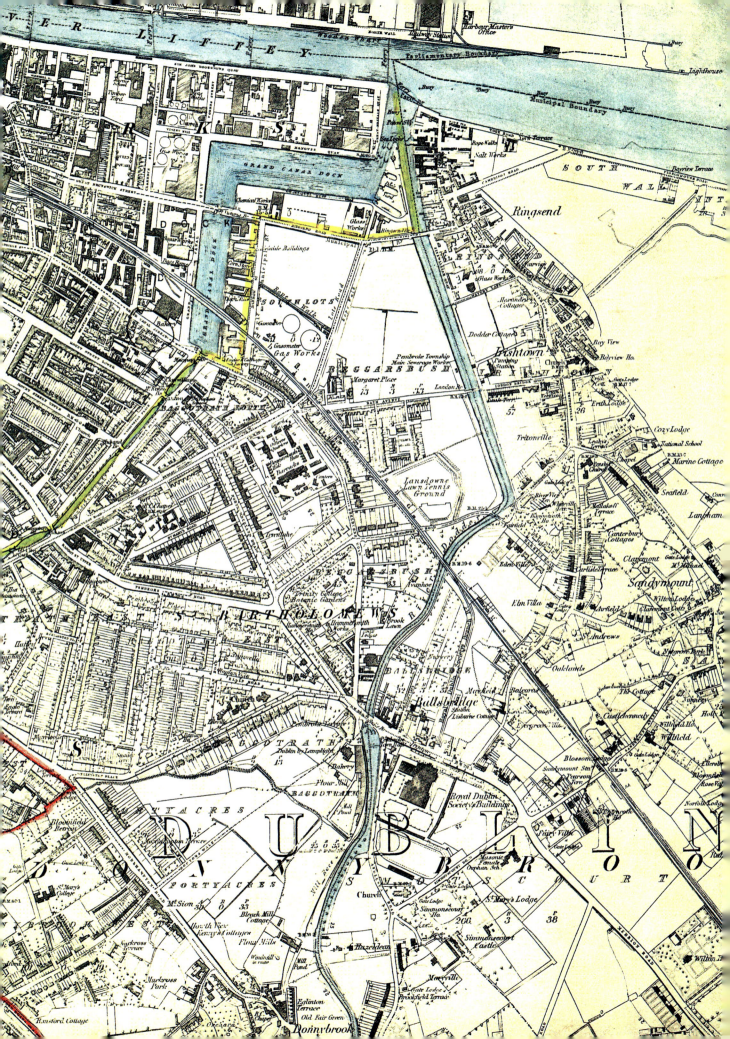

away from the immediate vicinity of the river there existed, in the nineteenth century, a leading engineering enterprise. The renowned Turner's Hammersmith Iron Works stood on the site of the later Veterinary College. Its most noteworthy products included the railings that surround Trinity College and the famous glasshouses that adorn the National Botanic Gardens, the Palm House at Kew Gardens and similar amenities elsewhere.

Other important centres of employment in the former Urban District are noteworthy. One was the huge office complex belonging to Irish Hospitals Sweepstakes which flourished from the 1930s to the 1970s, providing work for several hundred women. Before that, the site was occupied by Ramsey's Nursery. A large apartment complex, which takes its name from the former sweepstake business now stands there.

An internationally famous institution in Ballsbridge is the Royal Dublin Society, where the Dublin Horse Show and a great variety of other high profile events take place. The RDS leased land from the Pembroke Estate in 1879 and the first Spring Show took place from 19th to 22nd April 1881, followed by the first Horse Show. Ballsbridge became the headquarters of the RDS when the Society moved from Leinster House in the 1920s.

From 1879, Dublin Southern District Tramways horse cars ran to and from Blackrock through Ballsbridge. But passengers travelling to or from the city centre had to change cars at Haddington Road, the line from there to Nelson Pillar being operated by the Dublin United Tramways Company. On 16th May 1896 the Southern District Company opened Dublin's first electric service between Haddington Road and Dalkey. The Southern company was bought out by the Dublin United in September 1896, the electric cars reaching Nelson Pillar in 1898. The route through Ballsbridge, known as the Premier Line, was the last city tram service to close on 9th July 1949. Apart from passenger services, Ballsbridge was a very important tramway engineering centre.

With Sandymount and Donnybrook, Ballsbridge was one of three substantial shopping and business centres in the urban district.

RINGSEND/IRISHTOWN, although now a suburb of Dublin, is arguably the most important former village adjoining the Dodder. Its own local history and its relationship with Dublin are therefore worthy of more than a few brief references.

Dublin was originally a Viking city and after 1171, when an Anglo-Norman army seized it, Dublin became the centre of English rule in Ireland. The native Irish were therefore viewed as an alien force in the city. Suspicion of them was deepened by continual raids on Dublin and its environs by the O'Byrne and O'Toole clans from the nearby Wicklow Mountains.

By the 15th century, Irish migration to the city had made the English authorities fearful that English language and culture would become a minority there. As a result, the Irish inhabitants of Dublin were expelled from the city proper circa 1454, in line with the Statutes of Kilkenny. The Irish population were only allowed to trade inside the city limits by daylight. At the end of the day's trading they would leave and set up camp in what was to become known as Irishtown.

The origins of Ringsend and its first inhabitants come from the sea itself. It was originally called 'Rinn Aun', which literally means Sea Point. It has had many different names over the years such as 'An Rinn', meaning The Point. It has been named 'Raytown' which derives from the fact that it was a fishing village and today is called Ringsend, known as the Anglicised name for 'where the rings end'. O'Flanagan quotes a native who said "There were formerly rings along the quay wall, for the making fast (of) ships, terminating here, and this gave the place its name, Ringsend.".

Topographically separated from Dublin by the Dodder, up to the early nineteenth century Ringsend/Irishtown frequently experienced total isolation from the city because of severe flooding (both tidal and riverine) and the destruction of successive bridges over the river. Ringsend was an important industrial centre, details of which are described in Chapter 8 dealing with mills on the Dodder.

As can be seen on Thomas Phillips' map of 1685 MAP NO. 8, the port of Ringsend was built on a spit of sand ending at Ringsend Point on the River Liffey. Bernard de Gomme (1620-1685) was Chief Engineer to King Charles II, who reigned from 1661 to 1685. De Gomme and his assistant, Thomas Phillips, mapped Dublin during the 1670s. Their joint task was to inspect the fortifications of the capital and other places and to design fortresses for their better defence. As a result of the survey, plans for a massive citadel, a Fort-Royal, on the strand near Ringsend, were drawn up and shown on a map produced by de Gomme in 1673 (the proposed location is shown on MAP NO. 8) with a proposed causeway linking Lazy Hill to Ringsend. In 1684 Thomas Phillips was commissioned to survey fortifications in a number of towns and cities in Ireland. He again proposed that Dublin should be fortified with a citadel. The revised location would have been in the vicinity of Merrion Square as shown on MAP NO. 8. Neither the de Gomme nor the Phillips citadels were constructed.

J Cullen's map of 1693, MAP NO. 9, shows a road from Beggarsbush to Irishtown with a wooden bridge crossing the Dodder. This is now Lansdowne Road. The cross roads at Beggarsbush was located in the vicinity of the present day junction of Lansdowne Road and Shelbourne Road. .

J Cullen's map of 1706, MAP NO. 10, illustrates the extensive developments taking place in Ringsend at the beginning of the eighteenth century. The wooden bridge at Irishtown was still in use.

Edward Cullen's map of 1731, MAP NO. 11, details the Dodder estuary from Ballsbridge to Ringsend. At this time a bridge across the Dodder had been constructed at Ringsend replacing the wooden bridge at Irishtown and making the wooden bridge at Irishtown redundant. Cullen proposed a realignment of the river from Ballsbridge to Ringsend.

This was the bridge that connected Ringsend with the end of present day Pearse Street, originally known as Lazar's Hill and renamed Brunswick Street in 1796. It was recalled that "when the Grand Canal Company had built their floating and graving docks at the termination of Lazar's Hill, communication should be kept up, and neither transit by land or water interfered with."

For several centuries the marine terminal for passengers travelling between eastern Ireland and British ports was at Ringsend, and it was here

AN EXACT SUR- VEY OF THE CIT- ETY OF DUBLIN AND PART OF THE HAR- BOUR ANNO 1685

Deer Park

Kilmainham

The Hospital

DUBLIN

Citadel

Lazy Hill

Mill Island

Strand

Boney Broke River

Stone Bridge

Rings End

Strand

Irish Towne

Clontarf Island

Clontarf Head

MAP NO. 8 (OPPOSITE)
Map of Dublin & Liffey estuary
(Thomas Phillips, 1685)
showing proposed location
of citadel (at present day
Merrion Square)

MAP NO. 9 (RIGHT)
Map of the Strand between
Ringsend and Lazihill
(James Cullen, 1692)

MAP NO. 10 (BELOW)
Map of proposed
expansion of Ringsend
(James Cullen, 1706)

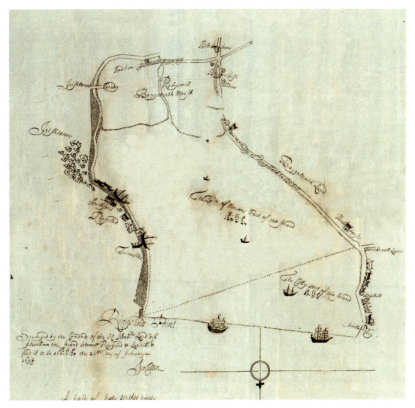

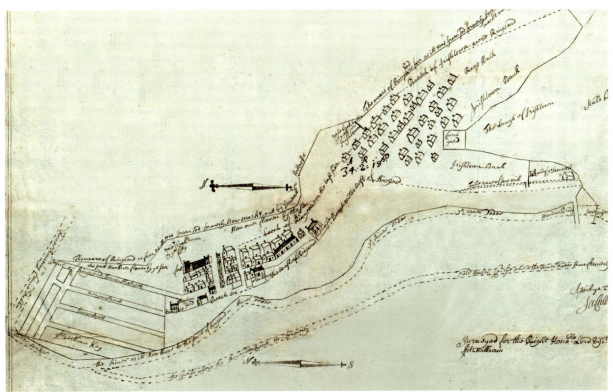

that Cromwell landed in August 1649. Larger vessels, unable to navigate further up the shallow and not yet embanked Liffey, also tied up at Ringsend.

Ringsend enjoyed considerable prosperity during the seventeenth and early eighteenth centuries. It was the port for mail packets (postal ships) from 1796 until 1818, but lost out to Howth, which could accommodate larger vessels. Unable to accommodate the steamships which constantly increased in size and numbers during the nineteenth century, Ringsend went into a long period of decline and worsening poverty which was not reversed for many years, when land reclamation, industrialisation and new roads gradually revived the area. The engraving, by Giles King from a drawing by William Jones, in **FIG. 2** shows Ringsend and its port in the eighteenth century

In the late eighteenth century, two major engineering works were carried out in the Ringsend area—the South Bull Wall and the Grand Canal Basin. Both projects provided expertise which was to prove extremely useful in solving the problems surrounding the building of the present Ringsend Bridge which are recounted in the section on Dodder bridges in Chapter 7 (Bridge No. 37).

Engineer William Jessop (1745-1814), built the Grand Canal through very boggy areas in the midlands, thereby becoming expert in solving construction difficulties encountered in soft ground conditions. However, building the entrance locks to the Grand Canal posed a much larger challenge. Forces imposed by the tidal zone, the loading conditions of filling and emptying locks

FIG. 2

A view of Howth and Dublin Bay from Beggars Bush (1745).

1. Ringsend; 2. Irishtown;
3. The South Wall;
4. The North Wall;
5. Clontarf Island; 6. Poolbeg;
7. Sheds of Clontarf;
8. Contarf; 9. Howth;
10. Lord Howth's House;
11. Maiden Tower;
12. Bailey Lighthouse;
13. Ireland's Eye;
14. Dublin Bay.

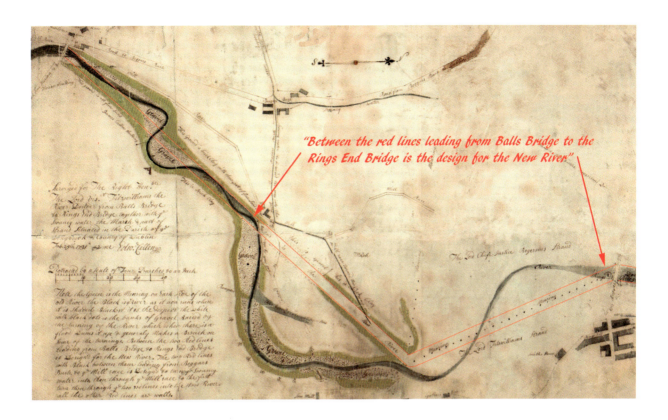

"Between the red lines leading from Balls Bridge to the Rings End Bridge is the design for the New River"

MAP NO. 11

Map of proposed
Dodder realignment
(Edward Cullen, 1731)

and the provision of cofferdams to enable their construction under water were unique and not encountered on such a scale elsewhere in Britain or Ireland.

Jessop completed the basin in 1796 and resolved the issue of foundations for the locks using the principle of the "inverted arch" to transfer the loads from the water and the walls into the ground and also reducing the depth of the foundation. In addition, the inverted arch acts as a restraint on the side walls of a lock so that they do not collapse when the water level is lowered. The term "inverted arch" describes the fact that it is the reverse position to the accepted norm, that is, over a door or supporting a bridge.

In Jonathan Barker's map of 1762 [MAP NO. 12] the street pattern of Ringsend and Irishtown, as we know it today, was well established. Between Ringsend and Irishtown there is a large pond known as "*My Lords Pond*", which was owned by Lord Fitzwilliam. Sea water entered the pond and was allowed to evaporate, leaving a residue of salt. Some houses were built here in the eighteenth century. In the early part of the twentieth century part of the foreshore had been reclaimed and used as a playground. The locals referred to this area as "My Lord's Park". It is now a public park. Reverting briefly to salt, this necessity was processed from sea water in two Ringsend factories in the eighteenth century.

The westerly basin of Jessop's Grand Canal Docks effectively extended Ringsend towards the city. Opened in 1796, the whole complex impressed John Ferrar who wrote: "Riding to Ringsend we were presented with a striking proof

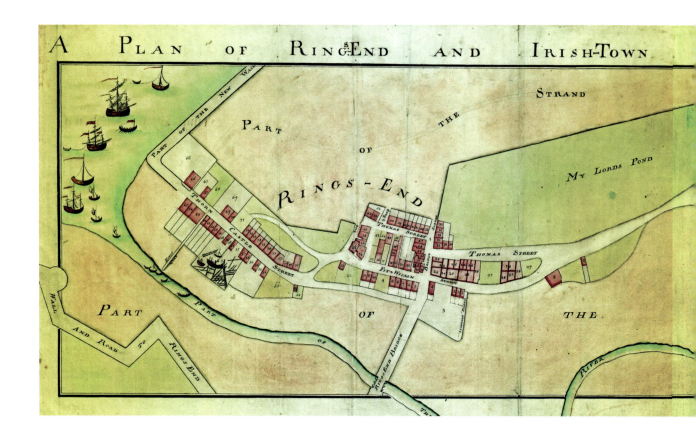

of the vast extent of human labour and human genius in docks building there; and we were highly pleased to find Councillor Vavasour retracting a great tract of waste ground near the bridge (over the Dodder)."

Few bridges have been as well documented as the one crossing the Grand Canal Dock from Ringsend to Pearse Street and its history is worth recording in some detail. In November 1879 The Irish Builder published details of what was known as the Brunswick Bascule, formerly called the Ringsend Draw Bridge. This was the bridge that connected Ringsend with the end of present day Pearse Street, originally known as Lazar's Hill and renamed Brunswick Street in 1796. It was recalled that "when the Grand Canal Company had had built their floating and graving docks at the termination of Lazar's Hill, communication should be kept up, and neither transit by land or water interfered with."

A description of the bridge was given by Rev. CT McCreary in Dublin Street Names in the 1890s. The timber draw bridge, dating from 1791, had a narrow opening section. The Irish Builder stated that "the timber draw bridge, or rather lifting bridge—more properly called Bascule—was the best the science of the day could provide. Generally known as the Brunswick Bascule, its approaches posed problems for vehicles negotiating it in wet weather. McCreary records that it was replaced in 1867 by a metal bridge named after Queen Victoria.

Ringsend never had horse trams, but an electric service connecting Sandymount with the Queen Victoria Bridge opened on 4th July 1900.

MAP NO. 12

Plan of Ringsend & Irishtown

(Jonathan Barker, 1762)

The Dublin United Tramways Company had a new Victoria Bridge erected to facilitate through traffic. It consisted of two swivelling structures side by side, the lines divided by parapets. The ironwork was by Ross and Walpole, whose foundry was located on the North Wall. Following completion of the bridge, the tram service was extended to the City Centre on 8th Match 1901. A tragic accident occurred at Victoria Bridge on 29th June 1904 when two little brothers, Joseph (4) and Patrick (3) Ward, were thrown out of their pram at the kerb. They were struck by the fender of tramcar No. 89 driven by William Ryan and died on the spot.

A most unusual collision took place in February 1928 when the Arklow schooner *Cymric* was waiting to enter the Grand Canal Dock at Victoria Bridge. The vessel was blown by a gust of wind and the bowsprit struck tramcar No. 223 which was crossing the bridge; there were no injuries. Unfortunately, the *Cymric* was lost in 1944 with the loss of all hands.

One of the most famous ships in Irish history, the Ouzel galley, sailed from Ringsend on a trading mission in the autumn of 1695. When it did not return within the expected period, it was assumed to have been lost with all hands. A few years later, the Ouzel reappeared, laden with riches, the origin of which were suspicious. Some of the sailors' wives were said to have remarried and had second families. As a result, children of questionable parentage were known in Ringsend as Ouzelers.

Victoria Bridge became an increasingly dangerous obstacle to traffic in later years. The tram service gave way to buses in 1940 and the opening of Ringsend bus garage in 1941 resulted in very frequent crossings of the bridge. Replacement of the bridge took place as soon as circumstances allowed but the opening ceremony for the new McMahon Bridge did not take place until 29th May 2008. This is the fifth bridge at this site in 217 years.

Before the domestic water supply, which we take so much for granted became commonplace, Ringsend provided an important amenity for Dubliners. At a time when personal cleanliness and hygiene were difficult to maintain Professor John de Courcy in *Anna Liffey – The River of Dublin,* points out that the Rocque-Scale map of Dublin published in 1773 shows a "bath for men" at Ringsend, and some 300 metres along the shore nearer to Irishtown, a "bath for women." In 1818, it was reported that 20,000 people bathed at every tide in Dublin Bay, some for pleasure and all for "the preservation of health". This was long before any but a few very select houses enjoyed indoor plumbing or bathrooms. MAPS NOS. 6 & 7 of Ringsend illustrate the large developments that took place in this part of the city during the second half of the nineteenth century.

Part of Ringsend fell within the area of Pembroke Township (later Urban District), all being incorporated into the City in 1930. And despite all the changes it has undergone, Ringsend has managed to retain the recognisable features and distinctive atmosphere of a village. Since the 1990s, much redevelopment has taken place, increasing its population considerably.

Many Ringsend people can trace their family trees proudly through several generations that were born and lived in the old village. Details of

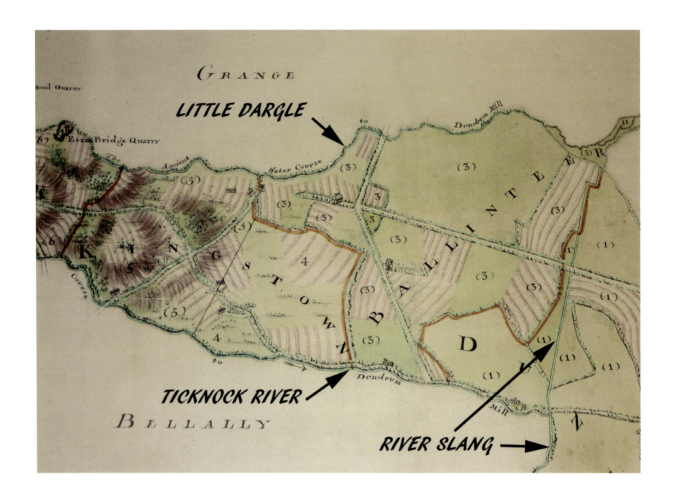

MAP NO. 13
Map of ancient water courses
(Little Dargle, Slang & Ticknock
Rivers) to Dundrum mill
(Johathan Barker, 1762)

the appalling poverty and suffering endured by those who eked out a mid-nineteenth century existence have been well documented by, among others, Mother Mary Aikenhead.

For many years, Dublin Corporation operated a ferry service which connected Ringsend across the Dodder to Sir John Rogerson's Quay and across the Liffey to North Wall Quay. When the East Link Bridge opened in 1984 the ferry service was discontinued. Ringsend Bridge, which has a long history, is described in the section on Dodder Bridges in Chapter 7 (Bridge No. 37).

DODDER TRIBUTARIES

As well as some minor streams and rivulets, the Dodder has a number of more substantial tributaries [MAP NO. 5]. Travelling downstream (West-East), the Dodder's principal tributaries flowing from the Dublin Mountains are listed below.

1 THE TALLAGHT STREAM 8.2km (5 miles) long, rises at Knockand, flows west, north and eastwards through Tallaght to join the Dodder near Firhouse. Its catchment area, now substantially in a built-up district, is 12.9 square kilometres

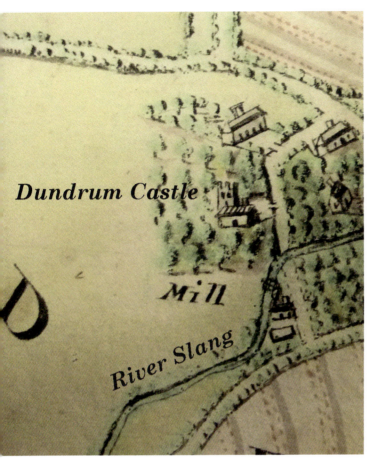

(5 square miles). The Tallaght Stream flooded the N81 Blessington road during the storm of 24th October 2011.

2 THE OWENDOHER, also known as the **Rathfarnham River** and the **Cruagh River**, as stated by JR O'Flanagan in series of articles about the Dodder in *The Dublin Saturday Magazine,* published in the nineteenth century. It is 10km (6.2 miles) long and is formed of two streams, Killakee Stream (from the slopes of Killakee Mountain); and Glendoo Stream (from west of Glendoo Mountain). These branches merge near Rockbrook Cemetery, and the river flows north to Ballyboden and on to Rathfarnham.

A major tributary, the **Whitechurch Stream,** also known as **Glin Stream**, rises near Tibradden, flows northwards through Grange Golf Club and St. Enda's and joins the Owendoher at Willbrook. The Whitechurch Stream is 8km (5 miles long) and has a catchment of 8.3 sq. kilometres (3.4 sq. miles). Serious flooding took place on this stream in 2007.

The Owendoher joins the River Dodder south of Bushy Park near the former settlement of Butterfield, just south-west of Rathfarnham Bridge. The Owendoher drains 21.2 sq. km (5,230 acres).

3 THE LITTLE DARGLE RIVER is 8km (5 miles) long, and drains an area of 10.05 sq. km (2,483 acres). It rises above Ticknock and joins the Dodder 400 metres upstream of Orwell Bridge. This river is culverted on its lower reaches and, just before its confluence with the Dodder, is joined by a rivulet known as the Castle Stream.

The Little Dargle River flows through Marlay Park. In the eighteenth century the banker David la Touche laid out the estate and built a threshing mill on the Little Dargle.

Where the present day river crosses Stonemasons Way (beside Hillview Estate) the flow was intercepted and diverted into the Slang catchment. On Jonathan Baker's 1762 map, of the Fitzwilliam Estate Lands of Dundrum, the upper section of the Little Dargle river is noted as "Ancient watercourse to the mill in Dundrum" [MAP NO. 13]. This would suggest that this section of the river was diverted to combine with the flow from Ticknock River to run the seventeenth century mill beside Dundrum Castle [MAP NO. 14]. This was probably a corn mill with a horizontal waterwheel. This diversion was the source of the Slang River. The flow was controlled by sluice gates with excess flow over and above that needed for the mill discharging back into the natural watercourse of the Little Dargle River. This arrangement is shown on Mallet's map [MAP NO. 4] and the 1912 Ordnance Survey 25 inch map. This arrangement gave the Dundrum Mill water rights to the Little Dargle.

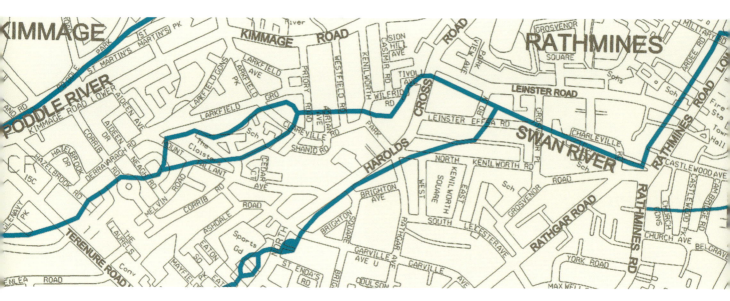

This facility to divert water from the Little Dargle catchment led to friction between the land owners in the catchment and the mill owners on the Slang. When there was low flow in the Slang the mill owners extracted most of the water flowing in the Little Dargle thereby depriving the land owners lower down the catchment of a water supply. The land owners would break the sluice gate controlling the flow which was then repaired by the mill owners. This happened many times in the nineteenth century.

The link to the Slang River was removed about 1980 and the river bed was realigned to its present course.

4 THE SLANG RIVER (*Abhann na Stéille* in Irish) is also known as the Dundrum River. The source of the Slang is the Meadowbrook, Ludfort and Hillview catchment area to the west of Ballinteer Road which was formerly a marsh. The river's Irish name may derive from *stiall* 'plank, board' (gen. *stéille*), possibly indicating that there may have been a walkway or panelling across the river.

In the National Archives, a coloured map [MAP NO. 13] in the Pembroke Estate papers shows the Fitzwilliam Estate lands in Dundrum, surveyed by Jonathan Baker in 1762. It shows a mill on the right bank of the River Slang near Dundrum Castle and two water courses to Dundrum Mill. The southern watercourse is the present Ticknock River feeding into the Slang River at Ballinteer.

The second watercourse on the map is called "The Ancient Water Course to Dundrum Mill". This is the present day Little Dargle, feeding into the Slang river. Baker's map suggests that the mill owner in Dundrum had water rights to the upper reaches of the Little Dargle. This connection to the Little Dargle is shown on Mallet's map.

The Ticknock River flows from the Three Rock Mountain down through Ticknock to Ballinteer where it joins the Slang River. The Slang then loops east,

MAP NO. 15

Present day Swan River
in underground culverts

north and west, passing through the Dundrum Town Centre retail complex; and finally flows northwards to the Dodder via Windy Arbour.

Flow from the Slang fed millraces and mill ponds at Dundrum, Windy Arbour and Milltown. The Slang is 8km (5 miles) long, draining 10.5 square kilometres (2,600 acres).

Immediately upstream of its confluence with the Dodder, there was, in the late nineteenth and early twentieth centuries, a series of septic tanks on the east bank of the Dundrum River. This installation, the property of Rathdown No. 1 Rural District Council, discharged effluent from the Dundrum area to the Dodder. In 1911, this outfall was diverted into a sewer laid along or near the southeast bank of the Dodder from Milltown to Ringsend.

THE SWAN RIVER – RATHMINES DRAINAGE

In the early nineteenth century, the multi-branched Swan River [MAP NO. 15], which rose near Kimmage, flowed north-eastwards through Mount Tallant and the then sparsely populated areas of Rathmines, Mount Pleasant Square, Ranelagh and Ballsbridge. The Swan and its five branches, totalling about 10 miles (17 kilometres) in length, originally went to sea through the Ringsend sloblands that had been partially reclaimed in the 1790s. In 1881, the Swan became a major component of the Rathmines and Pembroke (R & P) Joint Main Drainage Scheme, a combined system that carries both foul and surface water. At Clyde Road, an overflow on the R & P sewer allows storm water to discharge to what remains of the Swan River during heavy rain.

From Clyde Road, this short remnant of the Swan passes near Northumberland Road, Lansdowne Road and O'Connell Gardens. It enters the Dodder through a rectangular opening about 240 metres downstream of New Bridge (Lansdowne Road) and 140 metres south of Londonbridge Road. Urban mythology asserts that, many years ago, at least one resolute rugby enthusiast, lacking a match ticket, struggled up this culvert in the hope of finding a convenient manhole from which to emerge inside the Lansdowne Road Stadium.

Under and near Londonbridge over the Dodder and adjacent to the former pumping station, there is a multiplicity of important sewers. The oldest component of this complex is part of the 1881 Rathmines and Pembroke Main Drainage scheme. From Ballsbridge, the R & P No. 1 sewer flows northwards along the west bank of the Dodder and siphons under the river downstream of Londonbridge Road to continue eastwards to Ringsend. On each bank of the river, there is a siphon house, one of which was used as a morgue in former times.

No. 2 sewer, from Merrion, running parallel to the east bank of the Dodder, and No. 3 from Shelbourne Road, originally went to the Londonbridge Road pumping station, which propelled the flow into No. 1 sewer. In the mid-1980s, Nos. 2 and 3 were diverted into the new Grand Canal trunk sewer, which passes under the Dodder on the upstream or south side of Londonbridge Road. Diverting the two Rathmines & Pembroke sewers rendered the pumping station redundant, but what remains, including the brick chimney, has been restored as an important example of industrial archaeology and municipal history.

Another sewer flowing along the east bank of the Dodder is that from Dundrum, already mentioned in connection with the closure of the sewage tanks beside the confluence of the Dundrum River and the Dodder at Milltown. This metal conduit, known as the County sewer, continues northwards past Londonbridge Road to join the City Centre (Hawkins Street-Ringsend) trunk sewer at Fitzwilliam Quay.

The book produced by the Corporation to commemorate the opening of the city's main drainage system in 1906 describes the problems encountered in constructing the 8' 0" (2.43m) diameter sewer tunnel from Hawkins Street to Ringsend. Much of the line "is through ground composed of loose filling—sand, gravel, earth and is what is known as made ground." The work was carried out by the Greathead Shield process.

In the section beyond Victoria Bridge (site of the present McMahon Bridge) "even greater difficulties were met... In passing under the River Dodder there were only from two to three feet of clay between the top of the sewer and the water in the river, so that naturally there was great anxiety on the part of the contractors and engineers and all concerned until this portion of the work was executed, as provision had to be made for the escape of the men and engineers in case the river came in on them during the process of driving the shield; and indeed so porous was the ground that air bubbles could be seen coming up all over the river, close to where the shield was working and also in other places where there was water on the surface."

CHAPTER 5

THE PODDLE

s inferred earlier, when looking at the tributaries of the Dodder it is impossible to merely mention the Poddle. This legendary watercourse, the subject of innumerable articles and papers [MAP NO. 16, CITY WATERCOURSE AND PODDLE CATCHMENT – MALLET 1844], continues to fascinate school children and others given a river project to study -so much so that the Drainage Division staff have been known to call Spring the Poddle season. The Poddle rises at Fettercairn near Tallaght, has a nominal catchment of about five square miles (2.59 square kilometres) and even had some tiny tributaries. In the past, the Poddle was the antithesis of a tributary, having for hundreds of years taken water from the Dodder rather than augmenting it.

In *The Rivers of Dublin* (Dublin Corporation, 1991), C. L. Sweeney describes how the Poddle, called by at least eight different names in various epochs and areas, has played a major role in the social and economic growth of the City. With its long history of mystery and intrigue, quarrel and litigation, it features prominently in folklore and is a favourite subject for history and geography projects. Research into the Poddle's former roles as water supply, sewer and source of power for industry, offers a unique perspective on Dublin's past.

Among the alternative names for the Poddle quoted by Sweeney are the Puddle, the Cammuche or Kammouche (Crooked Water), and the Glascholach (at Harold's Cross). Its fifteen associated watercourses and branches include the Abbey Stream, the Old City Watercourse, the Glib Water, the Limerick Watercourse, the City Ditch, the Tenter Water, the Commons Water, the Hangman's Stream, the Camac Millrace, the Lea Brook or Crockers Barr Stream, the Lakelands Overflow and the surface water conduit of the Greater Dublin Drainage Scheme.

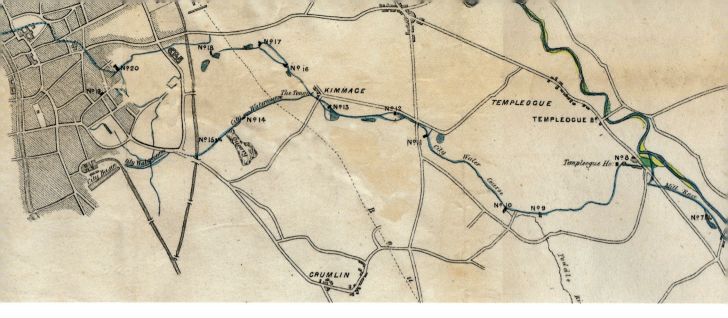

Other names bestowed on the Poddle included the Solagh, Sallagh and Sologh. These are all variations of the derogatory Irish word Salach (dirty), applied to the river during the period in which it was a putrid open sewer. A fourth noteworthy variation of this title occurs in *Down by the River Saile* (also called *Weila Weila Waille*), the song popularised by Ronnie Drew and the Dubliners in the 1960s.

The earliest Viking invaders are said to have moored their ships in a black pool (Dyflyn or Dubh Linn) on the Poddle, just upstream of its confluence with the Liffey [MAP NO. 17]. This pool, which gave the city its name, was where the garden behind Dublin Castle is today. The Vikings built a fort on the high ground overlooking the Poddle, which also supplied water to these first settlers. MAP NO. 18 indicates the development in Dublin between 1000 and 1300AD. In this map the Poddle is shown joining the Commons Water and flowing down both sides of the present day Partick Street, flowing around the walls of the city and discharging into the Pool which gave its name to Dubh Linn — Dublin.

Around 1300 six watermills are shown on the Poddle — Schyeclappe Mill, Pool Mill, two King's Mills and two Doubleday's Mills. The River Steine is shown flowing around St Stephen's Green and discharging into the River Liffey estuary. Thomas Phillip's map of Dublin in 1685 [MAP NO. 8] shows the development that has taken place since 1300. The estuary is being reclaimed with a river wall completed on the southside as far as present day Butt Bridge. The slobland to the north of Lazy Hill (later known as Lazar's Hill and now Pearse Street) is being reclaimed and built on. From the city Lazy Hill Road (present day Shelbourne Road) continued around the coastline as far as the stone bridge at Ballsbridge. Below Ballsbridge the Dodder was tidal and entered the Liffey estuary as shown on Thomas Phillip's map of 1685 [MAP NO.8].

The course of the Poddle in the distant past is not clearly mapped but it most likely followed its present one through Tymon and Kimmage as far as Captain's Road and then through the natural valley between Kimmage Road and Harold's Cross Road [MAP NO. 19: PODDLE FROM THE TONGUE TO RIVER LIFFEY]. It flows a little to the west of Leonard's Corner and then under Blackpitts, past St. Patrick's Cathedral, continuing via Ross Road. Entering the Castle Yard at the Ship Street Gate and flowing eastwards around the southern and eastern ramparts of the Castle, it passes the Bermingham and Record Towers [MAP NO. 20: RIVER PODDLE AT DUBLINCASTLE].

MAP NO. 16
City Water Course
& Poddle Catchment
(Robert Mallet, 1844)

MAP NO. 17 (OPPOSITE TOP)
Dublin circa 1107

MAP NO. 18 (OPPOSITE BOTTOM)
Dublin circa 1300 with locations
of medieval watermills

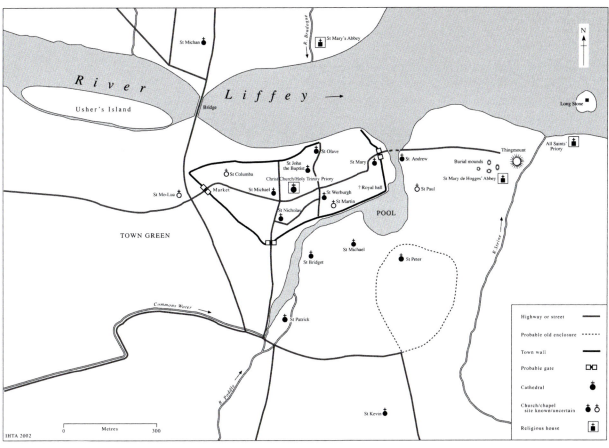

Map 1 labels:
St Michan, R. Broadmeadow, St Mary's Abbey, **River Liffey →**, Usher's Island, Bridge, Long Stone, All Saints' Priory, St Olave, St John the Baptist, St Mary, St Andrew, Burial mounds, Thingmount, St Columba, Christ Church/Holy Trinity Priory, St Mary de Hogges' Abbey, St Mo-Lua, Market, St Michael, St Werburgh, ? Royal hall, St Paul, St Nicholas, St Martin, POOL, TOWN GREEN, St Bridget, St Michael, St Peter, R. Steine, Commons Water, St Patrick, R. Poddle, St Kevin

IHTA 2002

Highway or street	————
Probable old enclosure	- - - -
Town wall	▬▬▬▬
Probable gate	▢▢
Cathedral	✠
Church/chapel site known/uncertain	✠ ○
Religious house	▣

Metres 0 — 300

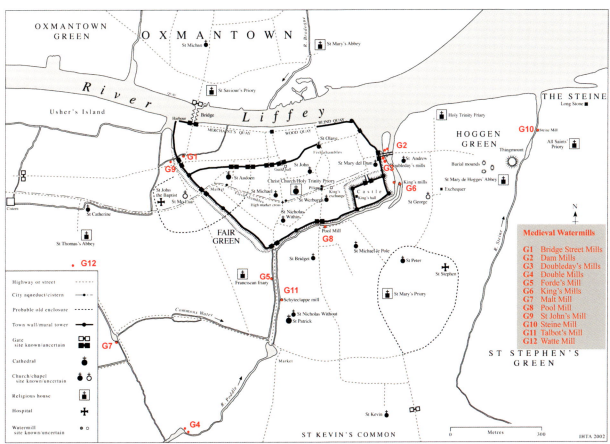

Map 2 labels:
OXMANTOWN GREEN, **OXMANTOWN**, R. Broadmeadow, St Michan, St Mary's Abbey, **River Liffey**, QUAY, St Saviour's Priory, THE STEINE, Usher's Island, Harbour, Bridge, BLIND QUAY, Holy Trinity Friary, Long Stone, G10 Steine Mill, HOGGEN GREEN, All Saints' Priory, MERCHANT'S QUAY, WOOD QUAY, St Olave, G2, Doubleday's mills, St Andrew, Thingmount, Fishshambles, St Mary del Dam, Burial mounds, G9, G1, St John, St Mary de Hogges' Abbey, Cistern, St John the Baptist, St Audoen, Christ Church/Holy Trinity Priory, Pillory, King's exchange, King's mills, G6, St George, Market, St Michael, Castle, Exchequer, St Mo-Lua, St Werburgh, King's hall, Fishshambles, High market cross, St Nicholas Within, St Catherine, FAIR GREEN, Pool Mill, G8, St Thomas' Abbey, St Bridget, St Michael le Pole, St Peter, St Stephen, G12, Franciscan friary, G5, St Mary's Priory, G11, Schyteclappe mill, Commons Water, St Nicholas Without, St Patrick, G7, Market, R. Poddle, G4, St Kevin, ST STEPHEN'S GREEN, ST KEVIN'S COMMON

IHTA 2002

Medieval Watermills

G1 Bridge Street Mills
G2 Dam Mills
G3 Doubleday's Mills
G4 Double Mills
G5 Forde's Mill
G6 King's Mills
G7 Malt Mill
G8 Pool Mill
G9 St John's Mill
G10 Steine Mill
G11 Talbot's Mill
G12 Watte Mill

Highway or street	----
City aqueduct/cistern	•—•
Probable old enclosure	- - - -
Town wall/mural tower	●—●
Gate site known/uncertain	▢▢ ▣▣
Cathedral	✠
Church/chapel site known/uncertain	✠ ○
Religious house	▣
Hospital	✚
Watermill site known/uncertain	• ○

Metres 0 — 300

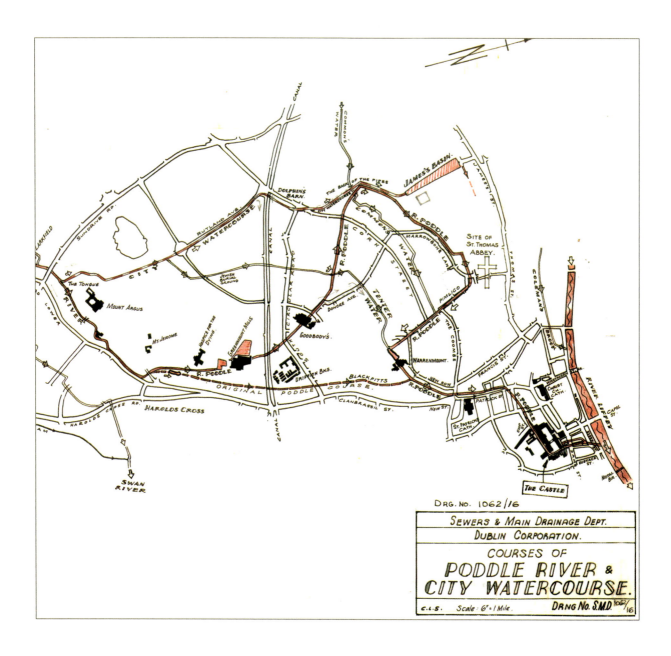

DRG. NO. 1062/16

SEWERS & MAIN DRAINAGE DEPT.

DUBLIN CORPORATION.

COURSES OF

PODDLE RIVER &

CITY WATERCOURSE.

C.L.S.   Scale: 6" = 1 Mile.   DRNG No. S.M.D. 1062/16

Turning northwards opposite the Chapel Royal, the Poddle bifurcates inside the Palace Street Gate, the more westerly branch flowing under the open space beside City Hall before going under Dame Street. The eastern branch flows under Palace Street, beside a bank, which a gang of would-be robbers tried to access from the culvert in the 1980s. Flowing under Dame Street and the Olympia Theatre, the main Poddle culvert is rejoined in Essex Street by its western or City Hall branch. It finally discharges into the Liffey through an iron grid or portcullis at Wellington Quay, outside the Clarence Hotel.

Sweeney records how enterprising citizens routinely regulated, diverted and abused the Poddle to serve local or personal purposes. Over the years, as

MAP NO. 19

River Poddle and
City Watercourse from
The Tongue to River Liffey
(circa 1800)

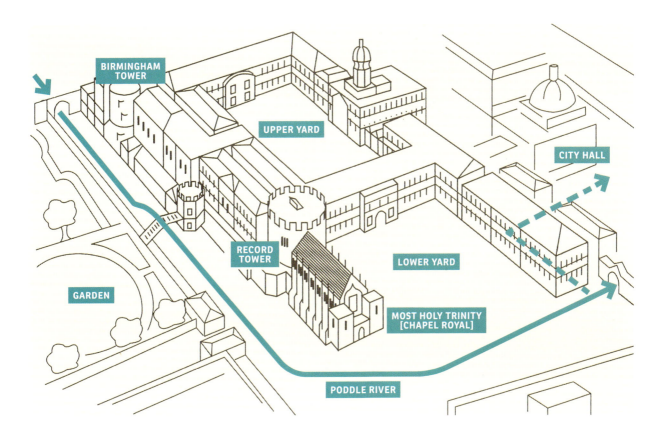

BIRMINGHAM TOWER

UPPER YARD

CITY HALL

RECORD TOWER

LOWER YARD

GARDEN

MOST HOLY TRINITY
[CHAPEL ROYAL]

PODDLE RIVER

MAP NO. 20

River Poddle at Dublin Castle

various alterations were made to these watercourses, several were abandoned or lay inactive until floods occurred. Because most of its water came from another river—the Dodder—and so many of its courses were man-made diversions, the Poddle has been described as an artificial river. Its dependence on the Dodder is recognised in several reports and records as, for example, in 1595 when a payment of thirteen pounds and two shillings was still due to the mason who had repaired "the water course at Dodder."

In heavy rainstorms, the Poddle flooded large settled areas in trying to revert to its natural course or bypass some artificial obstruction. During the late medieval period, disease-bearing water regularly inundated the Patrick Street district. The Poddle can still cause flooding, but the water is now relatively clean and better controlled than in the past. However, serious flooding occurred in the Harold's Cross and Lower Kimmage Road areas as recently as 1987, 1990 and 2011.

The expanding town of Dubh Linn became better regulated once the Normans established themselves and initiated some basic urban development. Dublin received its first charter from King Henry II in 1172, but in the sixteenth century it was still a small city of 46 acres surrounded by a wall with 45 towers. The maximum population within those walls has been calculated at 8,000, or 174 persons per acre.

FIG. 3
City weir at Balrothery
(Don McEntee)

Development of modern Dublin began following the restoration of the monarchy in 1660 and the appointment of James Butler, Duke of Ormond, as Lord Lieutenant. Around 1682 the population was estimated at nearly 60,000, which doubled to 120,000 by 1722. When the Act of Union came into effect in 1801, the population of Dublin, second city of the British Empire, stood at 172,000. By 1899, shortly before a major boundary extension, the City, then largely within the canal and circular roads cordon, was reckoned to contain 260,000 people.

Reverting to the twelfth century, when the first Norman bishop, Sean Coimin, built a cathedral outside the City in honour of St. Patrick, King John granted him the lands east of the Poddle and southwards to Harold's Cross. Around the same time, in 1176, a large monastery was built in honour of St. Thomas a Beckett on a site which is now part of Thomas Street, and the King granted the lands west of the Poddle and as far as Kimmage to this monastery. The river then formed the boundary between the lands of the Archbishop and those of the Monastery, which, being outside the City, administered its own laws.

The monks of the Abbey of St. Thomas, who constituted a powerful and influential body, are believed to have built an artificial branch of the river from the southern extremity of the Abbey lands near where the entrance to Mount Jerome Cemetery now stands. This new stream was wholly inside the Abbey lands and provided the monks with a convenient water supply. However, as the catchment basin of the Poddle River is less than thirteen square kilometres (five square miles) in extent, the dry weather flow at Mount Jerome would have been very small and probably shrank to a mere trickle towards the end of a dry summer.

FIG. 4

Start of city watercourse at Balrothery weir with the small opening through which the water entered the watercourse (Don McEntee)

Because the river could not be divided into two streams of any useful size and they dared not reduce with impunity the flow to the Archbishop's tenants, the monks realised that any additional water supply would have to come from some alternative source. Early in the thirteenth century they came up with a brilliant engineering solution to divert water from a weir on the River Dodder, which they owned at Balrothery, into an artificial watercourse to discharge into the Poddle.

By this time, the City had grown to a size at which an adequate and convenient water supply had become a matter of urgency. The citizens therefore voted money for this purpose and made representations to the King's most senior official in Ireland, the Justiciar Maurice Fitzgerald. In a writ issued on 29th April 1244, Fitzgerald commanded the Sheriff of Dublin, with the advice of the Mayor and citizens, to appoint without delay twelve freemen as jurors or inquisitors.

Their purpose was to identify a convenient source of water and how to conduct it to the City. At that time, Dublin lay largely to the west of the present Castle, around where High Street is now. The urgency of the project was emphasised by Maurice Fitzgerald's injunction that anyone who opposed the work was to be suppressed by force and brought before him at the next Assizes.

An agreement was made with the Priory of St. Thomas, which owned the weir on the River Dodder at Balrothery. This weir [FIG. 3] raised the level of the river high enough to permit the diversion of water through a man-made channel [FIG. 4] 1.75 miles (3km) long. The quantity to be drawn off at that time was limited to what could pass through the axle ope of a wagon. The Balrothery

FIG. 5
Sluice gate controlling flow
of water in city watercourse
Excess water flowed back
to Dodder through channel
on right (Don McEntee)

canal, later controlled by sluice gates [FIG. 5], left the Dodder in a north easterly direction, joined the Tymon River near the site of the future Mountdown Mills, and flowed from there to the City as the River Poddle.

Writing in 1879, Handcock stated that in 1456 John Pylle of Templeogue was sworn to "keep the water", i.e. to ensure a clean and adequate supply and bring it as far as the city cistern. In 1491 a man called Walsh was to have conduct of the water from the head at the Dodder to the tongue and harbour and from the tongue to the cistern was "to be kept as of old time". Handcock noted that the in earlier times it was easy to cross the river below the Balrothery weir; it was in fact a feature of the road from Greenhills to Tallaght. For many years the only way of crossing from Firhouse during a flood was to wade—dangerously—along the top of the weir.

At the Tongue Field, north of the present Sundrive Road, the flow in the river was divided by a masonry structure known as the Tongue [FIG. 6] to apportion and divert one third of the flow as a water supply to the Mayor's citizens. Under the agreement with the Priory, a down payment of one mark and an annual rent of five marks were to be paid to the monks. In 1259 an inquiry found that the city had drawn twice its allocation of water but had failed to pay any rent. Bills for improvements carried out on the Balrothery weir in 1555 also remained unpaid seven years later.

Maintaining the watercourse, especially the head works, was a formidable task. When inundations of the Dodder, notorious for its treacherous flash floods, caused damage, the Mayor and bailiffs were empowered to impress a number

FIG. 6

The Tongue (Stone Boat) on
River Poddle at Kimmage.
This divided the flow in Poddle
with one third of flow going
left to City Basin
(Don McEntee)

of citizens, as well as residents of the abbeys and monasteries, to repair the
damage. The Tongue was rebuilt in 1555 and its concrete successor, also known
as the Stone Boat, still divides the stream proportionately.

The Abbey of St. Thomas jealously guarded its ancient rights over the
Poddle until 1538, when King Henry VIII suppressed the monasteries. St.
Thomas's was then bestowed, with all its privileges, on Sir William Brabazon,
later the Earl of Meath. The section of the Poddle within the manors of
Thomas Court and Donore, also called the Liberties, henceforth became
known as the Earl of Meath's Watercourse. The Abbey authorities and their
successors, the Earls of Meath, enjoyed a monopoly over water supply and
charges within the Liberties until 1864, when this right was purchased by the
Corporation for £6,400.

THE CITY WATERCOURSE

The channel into which the one-third portion of the Poddle water diverted for
the City at the Tongue Field was known as the City Watercourse [MAP NO. 19]. It
flowed at a higher level than the Earl of Meath's Watercourse, via the present
Rutland Avenue and Dolphin's Barn Fire Station, under the Grand Canal (from
1756) and on through Dolphin's Barn to James's Basin.

A water storage cistern or basin (reservoir) was constructed on the site
where Basin Street is today, off James's Street. It took nearly ten years to
supply the majority of the citizens from the cistern. The earliest distribution
system consisted of open channels along the sides of Thomas Street and

High Street. In 1308, a fountain was erected in Cornmarket to supply poorer citizens. To provide the meagre daily allowance of six gallons of water to a population of 140,000, the City Watercourse would have to deliver a minimum of 840,000 gallons every day.

Instructions were issued on 18th November 1245 to have a supply laid on to the King's Hall at Dublin Castle. A length of lead pipe excavated in Castle Street in 1787 may have been part of this conduit laid down more than 500 years earlier by order of King Henry III.

During the sixteenth century, Dublin's public buildings, structures and community services suffered physical decay, ascribed to a scarcity of communal funding. The City Watercourse was said to be in constant need of repair; and it was well into the next century before the arrears of maintenance were overtaken and improvements carried out.

During its primacy in supplying water to Dublin, the City Watercourse and its extension into the city were blocked from time to time. According to Holinshed's Chronicles, in 1534 the followers of Silken Thomas "cut the pipes of the conduits whereby [the city] should be destitute of water." In 1597, the Talbot family of Templeogue stopped the supply in pursuance of a dispute with the owners of mills powered by the river. And, during the 1649 Cromwellian occupation of Dublin, Royalists blocked the Poddle, depriving the Parliamentary forces of both drinking water and power for their corn mills.

Lord Henry Barry, tried for murder in April 1739 and sentenced to death, was reprieved, exiled and later returned home. Accounts (possibly apocryphal) of these events state that Sir Compton Domville of Templeogue, through whose land the City Watercourse flowed, threatened to cut off the water supply if his nephew was executed.

The James's Street cistern, which served the more elevated parts of Dublin, was enlarged in 1671 and pipes—first of lead and later of timber—were laid from it to supply the city. In 1721, a new basin, covering 3.69 acres and capable of holding 9.5 million gallons or three months' supply, was built at James's Street. Because the new basin was ten to fifteen feet higher than its predecessor, the City Watercourse from Dolphin's Barn to the Basin was raised by embankments and masonry on a new easier gradient. From the basin, a ten-inch lead delivery pipe was laid into James's Street to connect with three six-inch mains leading to the City. This work was carried out under the supervision of the Surveyor General, Captain (Later Col) Burgh.

The flow in the Poddle and the City Watercourse fluctuated seasonally and was also adversely affected by mill owners and others who dammed the stream or abstracted water without considering the rights of others. Unscrupulous landowners outside the City habitually treated the Watercourse as their own personal property. In a 1735 pamphlet, now in the Library of Trinity College, architect Richard Cassells described the watercourse as being badly choked with weeds. There was seepage through the banks and in some places above Dolphin's Barn, the banks had been breached and the openings stopped with sods. With these obstructions removed, the breaches became open sluices allowing water to irrigate adjoining fields.

On occasion, regulations were introduced to ensure the quality of the water. Allowing animals to graze beside the open watercourses, dumping of any sort and washing of clothes were among numerous proscribed malpractices. In the case of offending animal owners, fines could be imposed at a fixed rate per beast for every day the nuisance continued. Preserving an uncontaminated supply relied heavily on the goodwill and common sense of the citizens, but dealing with stupid, irresponsible or malevolent individuals was always a problem — and it would obviously have been very difficult to police the Poddle and the City Watercourse constantly.

Industrial pollution became a serious problem as early as 1718, when people complained that the public water supply was unfit. The cause was traced to effluents from a tucking mill and a paper mill on the Poddle, the owners of which successfully argued that the Corporation had no powers to prevent the discharges. The Irish Parliament therefore enacted legislation in 1719 "for cleaning and repairing the watercourse from the River Dodder to the City of Dublin, and to prevent the diverting and corrupting the Water therein." The 1719 Act gave the Corporation much needed full control of the Poddle.

Further improvements were undertaken following the appointment, in 1735, of James Scanlon as Waterworks Engineer, a very early if not the first time this job title was used. The Poddle (City Watercourse) was Dublin's sole public source of water until 1745, when Scanlon erected a pump on the Liffey at Islandbridge. Driven by a water wheel, this installation delivered water to the expanding north side of the city.

Thirty years later, in 1775, water from the then new Grand Canal flowed into the City Basin, supplanting the Poddle as the city's primary source of water. In the years preceding the Canal supply, water was so scarce that it had to be rationed. The supply was turned on to the streets in rotation, and the turncocks rang a bell to warn the residents of each street when this was about to happen. The turncock was responsible for turning water valves on and off, thereby controlling the supply of water through the watermains. Dublin City Council still employs turncocks. An Act of 1780 required every premises supplied with water, whether trade or domestic, to have a cistern controlled by a stopcock or ball valve.

Around 1790, the Royal Canal superseded the Liffey to service the north side, via the Blessington Street Basin from 1806. The canals supplied Dublin with water until 1868, by which time the Vartry scheme was in full operation.

Even after it ceased to supply domestic water, the Poddle had to be maintained in the interests of the many industrial users along its courses. CL Sweeney estimated that some fifty mills and other industries depended on Poddle water over the years. Records are sketchy; some enterprises were short lived, others were started without notice and closed down without leaving any trace of their existence.

To maintain the supply, Andrew Coffey, Waterworks Engineer in the early nineteenth century, reconstructed the Balrothery weir, and the present sluice gates and by-pass channels date from that time. With the extraction of Dodder water no longer necessary, an overflow culvert was constructed from Kimmage in 1937 to take excess water from the Poddle back to the Dodder, serving Lakelands and Bushy Park en route.

Development since the 1970s all but obliterated the Balrothery Canal but, following Hurricane Charlie in August 1986, further major repairs were carried out and much needed restoration effected. This very historic engineering complex survives after nine centuries, perpetuating a link to the distant beginnings of our Dodder water supply.

Because they no longer had to supply domestic water, the Poddle and City Watercourse became increasingly contaminated with waste and sewage from the late eighteenth century onwards. Before Dublin's Main Drainage scheme became operational in 1906, the Poddle, flowing through densely populated poverty-stricken areas, became as great a nuisance as the Liffey into which it discharged. Conditions were appalling in its lower reaches and in the overcrowded dwellings of the Liberties, where filth accumulated in yards and streets.

In the early nineteenth century, the historians Whitelaw and Walsh wrote: "It [the Poddle] formed an immense sewer, carrying off the filth in its current, and putrefying the streets under which it passes. It occasionally, however, bursts from its caverns and inundates the vicinity to a considerable extent, particularly Patrick Street, Ship Street and the Castle Yard and Dame Street, where it is sometimes necessary to use a boat."

Writing later in the nineteenth century, William Handcock noted just how badly the Poddle had deteriorated over more than a hundred years following its replacement in 1775 by water from the Grand Canal as Dublin's source of potable water. Commenting on the diversion of Dodder water to the Poddle, he wrote "The citizens are fortunate that they are not now dependent on it, for it is so polluted by paper-making that it has become poisonous, and cattle and horses have died from drinking it."

The 1906 Main Drainage Scheme intercepted most of the foul matter that entered the Poddle within the city boundary, eventually resulting in the historic river becoming a surface water conduit. The largely redundant City Watercourse was culverted and diverted twice during the twentieth century, into the Grand Canal at Dolphin's Barn in 1931 and in 1984 to the Greater Dublin Drainage tunnel. This tunnel, built in the 1970s along the line of the Grand Canal, has separate compartments for foul and surface drainage and intercepts several surface water conduits, reducing the threat of flooding. Surface water from the tunnel discharges through the Grand Canal Dock from Maquay Bridge on Grand Canal Street to the Liffey.

Today, the Earl of Meath's watercourse largely follows its old route from the Tongue [FIG. 6] via Mount Jerome and Greenmount to Donore Avenue, where an ancient bridge has been built into the roadway.

Outside Donore Avenue Church, the old Poddle course is blocked off and a new culvert takes the flow along Donore Avenue to a point between Ebenezer Terrace and Harman Street to rejoin the old route. It then flows to the millpond at Warrenmount Mills before going under New Street and Patrick Street, which had a culvert at each side until about 1904. The abandoned western culvert was exposed and photographed during road widening in the 1990s and the area was thoroughly worked over by archaeologists.

THE PODDLE AS AN AMENITY AND A SECURITY RISK

As the City extended southwards during the twentieth century, the Poddle provided a convenient surface water drainage conduit for new developments. Several sections were culverted, mainly under new road crossings and some branches were abandoned or filled in. An interconnection to take surplus water from the Poddle to the Dodder catchment was completed in 1937. This flows through Terenure College and Bushy Park, where it used to feed the lakes and amenity streams.

A particularly attractive feature was created on the Poddle at Mount Argus in the 1980s, when a flood storage pond was constructed to reduce risks downstream. The river was redirected to its old course at this location in the 1990s and the pond was created as an amenity, complete with an ornamental fountain.

There are many stories of incidents, some of them highly questionable, involving prisoners who used the Poddle to escape from Dublin Castle. Such tales tended to become increasingly embroidered with each telling and there are always people eager to further exploit their romanticism. Ever since the original castle was built, the Poddle has been an obvious security risk. This risk is recognised when high profile personages are about to visit Dublin Castle. The underground section of the Poddle is checked and sometimes the manhole covers are welded.

The nerve centre of British Government power in Ireland until 1922, the Castle was especially vulnerable during the War of Independence. A report by George Harty, Engineer-in-Charge of the Sewers and Main Drainage Department during those terrible years, points this up. On 8th March 1921 he was taken, under duress, from his home to his office by Crown Forces in an armoured car. He was ordered to provide information regarding certain of the Ordnance Sheets and drawings showing the River Poddle and details of the sewers in the vicinity of the Castle. His office was thoroughly searched, and some plans and maps were taken, and he was coerced into accompanying a British army officer through a length of the Poddle culvert.

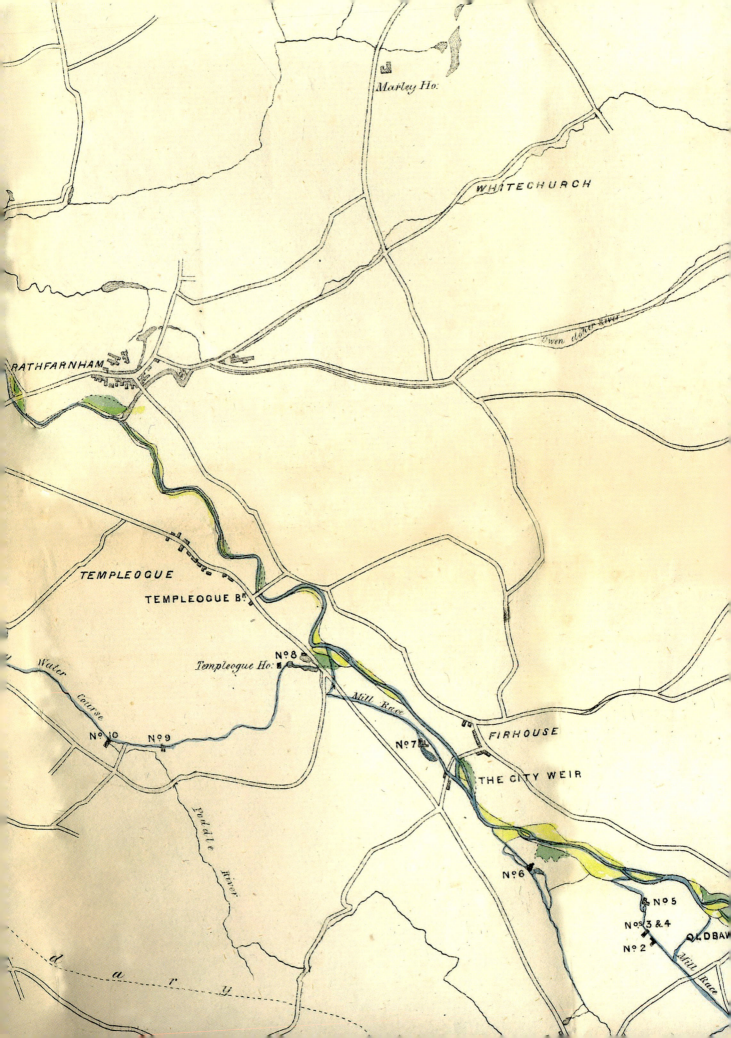

CHAPTER 6

THE DODDER FROM OLDBAWN TO RINGSEND

hile much of its course upstream of Oldbawn is in open country, the character of the Dodder changes markedly east of Oldbawn Bridge, [MAP NO. 21] exhibiting a wide range of characteristics. Flowing increasingly through a variety of suburban landscapes, the Dodder becomes the principal feature of several attractive linear parklands that change imperceptibly but constantly.

Between the Balrothery Weir and Rathfarnham Bridge, the Dodder follows a serpentine or looped course. Below Spawell, its course formerly covered a wide area that formed a strand or flood plain. The channel here was deepened, narrowed and straightened in 1846 and until 1912, stepping stones facilitated crossings at this location in fine weather. Some of the former strand now forms part of a riverside park.

Several more loops and bends bring the present day Dodder past Templeogue and on to Terenure, where it is an integral element of the amenities in Bushy Park and, east of Rathfarnham Bridge, an attractive roadside feature. As it flows on through Milltown, Clonskeagh and Donnybrook, [MAP NO. 22] some remnants of its industrial past enhance the river's surroundings before it disappears temporarily behind the houses on Anglesea Road.

Downstream of the weir at Ballsbridge for roughly one mile (1.6 km) above its mouth, the Dodder is tidal. "There can be little doubt that the low-lying parts of the Pembroke Township were once a lagoon, such as are now to be seen at Wexford and other places on the east and south coasts of Ireland. This lagoon, formed behind the raised beach on which Sandymount is built, may have had one or two outlets, and probably at one time the Dodder flowed out by Merrion." This quotation from Henry T. Crook's address to the Institution

MAP NO. 21

River Dodder from Old Bawn to Rathfarnham (Robert Mallet, 1844)

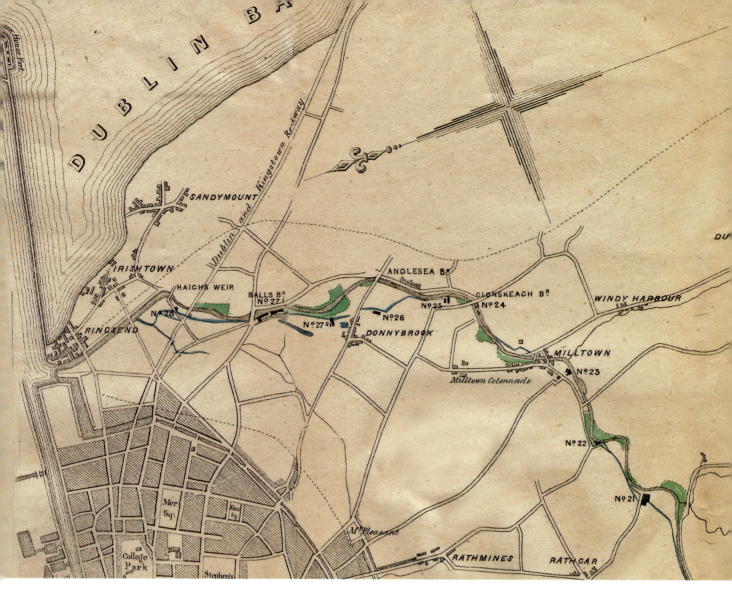

River Dodder from Milltown to
Ringsend (Robert Mallet, 1844)

of Mechanical Engineers in London on 1st December 1881 was one of several describing the Dodder's capricious and often treacherous conjunctions with the Liffey and the sea.

Gerard Boate, in his *Natural History of Ireland* (1652) wrote in some detail about the Dodder. Quoted in Weston St. John Joyce's *The Neighbourhood of Dublin,* Boate elaborated on the aftermath of serious flooding and the isolation of Ringsend bridge, built between 1629 and 1637: "Since that time a stone bridge hath been built over that brook upon the way betwixt Dublin and Ringsend; which was hardly accomplished when the brook in one of its furious risings, quite altered its channel for a good way, so as it did not pass under the bridge as before, but just before the foot of it, letting the same stand upon the dry land, and consequently making it altogether useless. In which perverse course it continued until perforce it was constrained to return to its old channel and to keep within the same."

The Dodder was also known as the Rathfarnham Water and the Donnybrook (or Donebrook) River in the seventeenth and eighteenth centuries. Before Sir John Rogerson's Quay and the South Wall were built, it was said to have flowed into the mouth of the Liffey near where the American Embassy now stands at Ballsbridge.

In *Anna Liffey – The River of Dublin,* Professor John de Courcy and Stephen Conlin noted: "In the seventeenth century, the Dodder split into a number of branches to form a delta from Ballsbridge to the Liffey. These flowed at first through meadows liable to occasional inundation and then, on about the line of Bath Avenue, spilled out into a strand that was covered twice a day by the tide." This is illustrated on MAPS NOS. 9, 10 & 11. On the west side of this delta was the old shore road from Ballsbridge to Dublin (Shelbourne Road/Grand Canal Street). By 1760, a large expanse of the delta was blocked off from the tides by Sir John Rogerson's Quay and an embankment along the line of what is now South Lotts Road.

"The east side was the slightly raised gravelly spit of land with Irishtown at its root and Ringsend at its tip [MAP NO. 12]. Like any delta with multiple branches, it was wayward and liable to change its course overnight. This it once did, cutting a new channel and leaving a new-made bridge high and dry with no water under it."

From time to time in the seventeenth century, various proposals were made for developments in the Dodder Delta. In his presidential address to the Institution of Civil Engineers of Ireland on 5th November 1945, Norman Chance instanced plans prepared in circa 1670s for a Citadel of Dublin where Merrion Square is today. This was a defence against the Dutch during the third of the four Anglo-Dutch wars fought in the seventeenth century. Among the notable engineers to whom this scheme was ascribed were Sir Bernard de Gomme and Andrew Yarranton. In de Gomme's map of 1673 the proposed citadel is shown encompassing Ringsend and a substantial part of the adjacent Dodder Estuary. In Phillips' map of 1683 [MAP NO. 8] the proposed citadel is shown on the west side of the estuary at present day Merrion Square. Yarranton also proposed locating the Harbour of Dublin in the Dodder Gulf.

For hundreds of years, the Dodder and its delta were a serious impediment to communication between Ringsend and Dublin City, which are only about two miles (3.2km) apart. Ringsend was the cross-channel passenger and packet harbour for Dublin until Howth took over this role in 1808. Meanwhile, convenient access between Dublin and Ringsend was at best intermittent. When the weather was dry and the tide was low, it was possible to ford the Dodder and cross the strand, but unfavourable conditions frequently necessitated a detour around Irishtown and Ballsbridge, where the bridge was often impassable.

Beyond what weather conditions might inflict, people travelling by Irishtown and Ballsbridge faced a further serious threat. This came from the highwaymen and footpads (muggers) who infested the area between the Dodder and Ballsbridge, which became known as Beggarsbush.

Until the last quarter of the eighteenth century, large quantities of material carried down by the Dodder into the Liffey worsened the already serious difficulties that plagued shipping. To eliminate this problem as well as that of bridging the river, it had been suggested in 1778 that the Dodder should be diverted through the neck of the peninsula near St. Matthew's Church in Irishtown to discharge on the sands of the South Bull. On Samuel Stoule's drawing for the Firzwilliam Estate in 1780 details of this diversion from

Ballsbridge to the sea are shown. Another diversion by Samuel Stoule details a route from Donnybrook discharging to the sea at Sandymount. If this diversions of the river had been done, Ringsend would effectively have become an island on which a fort could have been built to monitor ships on the Liffey and in the bay.

St. Matthew's (also known as the Royal Chapel) had been built in 1704 to cater for those worshippers in Ringsend who could not reach their parish church at Donnybrook in times of flood.

A report of January 1786 recorded that dredging had been carried out at the bar of the Dodder to deepen it and that 1,621 tons of ballast had been raised during two weeks in the previous month. Many of the problems associated with the confluence of the Dodder and Liffey were overcome when William Vavasour built the embankment walls on the lower reaches of the Dodder. To facilitate the construction of the Dodder canal Brian Siggins mentions that in 1790 the course of the river was diverted in a temporary channel along Newbridge Avenue discharging to the sea at Ringsend. Since the conclusion of this work in 1798, the river has flowed between these walls to discharge into the Liffey immediately upstream of where the Eastlink Bridge now spans it.

On 25th April 1972, some 73 metres of the embankment wall along Fitzwilliam Quay collapsed into the Dodder. Upon examination the entire length of this wall (173 metres) was found to be dangerous and was replaced. Another serious collapse occurred at Fitzwilliam Quay in 2011, which necessitated closing part of the roadway.

Completion of Vavasour's channel facilitated the reclamation of land that had previously been subject to flooding by the Dodder. But flooding along this section of the river could also be caused by occasional coincidences of certain tidal and climatic events. Occurrences of this nature in 1924 and 2002 are described later.

CHAPTER 7

FORDS, DEATHS
AND BRIDGES

n the early eighteenth century, there were fords on the lower Dodder at Milltown, Donnybrook, Ballsbridge and Ringsend. These crossings were always hazardous and virtually impossible to negotiate when the river was in spate, resulting in many drowning tragedies. Adjacent to where Milltown Bridge now crosses the Dodder, there was in the eighteenth century a ford that people used as a short cut to avoid going round by the somewhat inconvenient Packhorse Bridge.

The Milltown crossing was notoriously deceptive and dangerous, and several fatalities occurred there. In 1756, despite being warned, a man and a boy travelling on horseback to Powerscourt attempted the crossing but were carried away and drowned. The only daughter of the Clarke family was drowned here in 1781 and her father, steward at the House of Industry, was lost at the same spot a year later. The ascent from the ford was dangerously steep and in 1787 a child who was travelling on the roof of a mourning coach accompanying a funeral to Dundrum fell off and was killed. The foregoing are random examples of the many lives that were lost along the Dodder through the centuries.

Beatrice Doran records in her book on the History of Donnybrook that in the seventeenth century a bridge was built across the Dodder known as The Bridge of Simmonscourt. This was in the vicinity of Simmonscourt Castle. It was in bad condition by 1640 and Dublin Corporation voted to spend £10 on its restoration. During the 1641 rebellion the lands of Simmonscourt were laid waste. The bridge probably disappeared after this.

Until 1798, there were no bridges on the river upstream of Rathfarnham, but there were fords at Templeogue, Firhouse and Oldbawn. Patrick Healy records that around 1800, a man named Murphy attempted to negotiate the Oldbawn ford at night. The river, which was in flood, swept him away with his horse and cart.

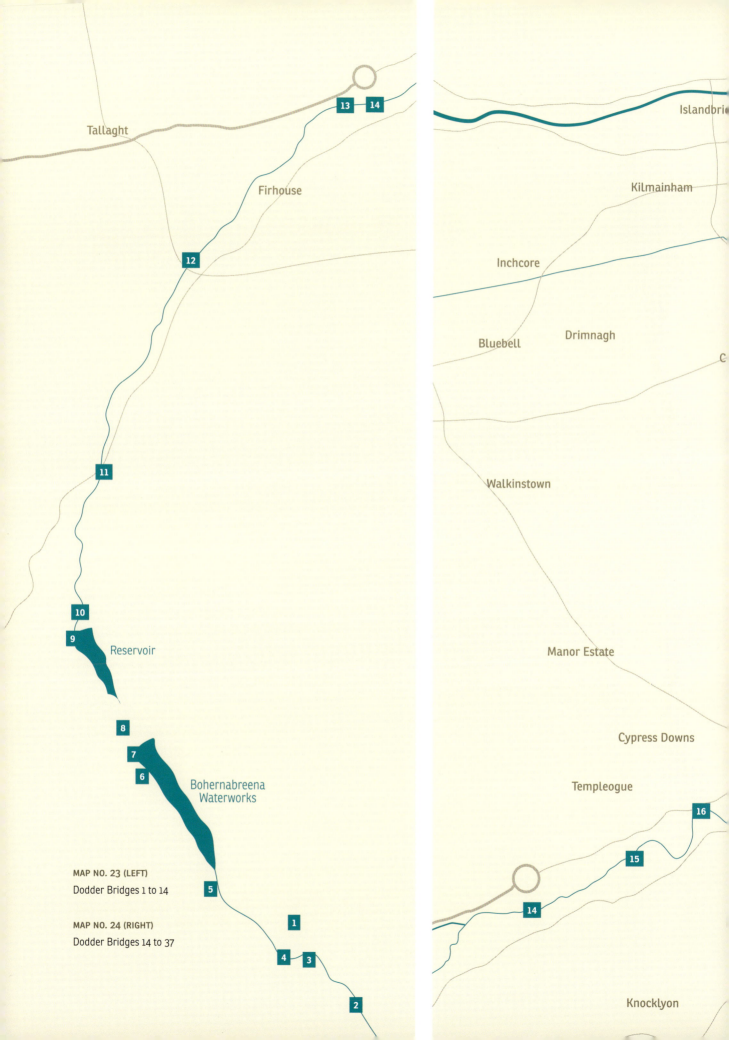

Tallaght

Firhouse

13 **14**

12

11

10

9 Reservoir

8

7

6 Bohernabreena
Waterworks

MAP NO. 23 (LEFT)
Dodder Bridges 1 to 14

MAP NO. 24 (RIGHT)
Dodder Bridges 14 to 37

5

1

4 **3**

2

Islandbric

Kilmainham

Inchcore

Drimnagh

Bluebell

C

Walkinstown

Manor Estate

Cypress Downs

Templeogue

16

15

14

Knocklyon

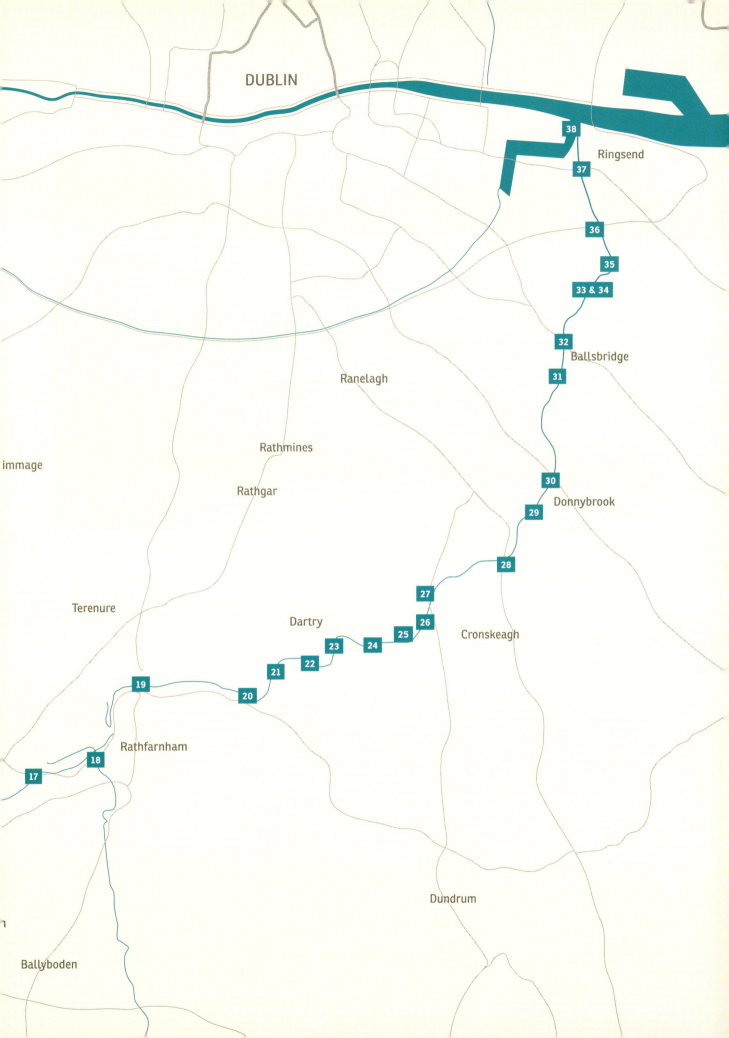

DODDER BRIDGES

From west to east the 37 bridges over the Dodder are shown on **MAPS NOS. 23 & 24**. At the upper end of the river, some of these see little use, while others are elements of the Bohernabreena reservoirs complex; the third group are major road bridges. Public bridges upstream of Bushy Park are in the area of South Dublin County Council, while those between Bushy Park and Clonskeagh connect Dublin City with the South Dublin or Dun Laoghaire Rathdown areas. Bridges from Clonskeagh seawards are within Dublin City.

NO. 1 CASTLEKELLY OLD BRIDGE. This bridge is the oldest one in the valley and crosses the old bed of the River Dodder. There is now only a small stream flowing under this bridge.

NO. 2 The most remote or uppermost crossing over the Dodder, southwest of Glenasmole and above the waterfalls at the first Castlekelly Bridge. It is a simple footbridge allowing access between fields on either side of the river.

NO. 3 CASTLEKELLY BRIDGE A. crosses the new channel of the River Dodder [FIG. 7] built circa 1885.

NO. 4 CASTLEKELLY BRIDGE B. crosses the new channel of the River Dodder built circa 1885.

FIG. 7
Castlekelly Bridge on new Dodder channel (Don McEntee)

NO. 5 CASTLEKELLY NEW BRIDGE. This bridge was destroyed by the flood of 25th August 1905. The bridge [FIG. 131] was rebuilt in February 1906.

NO. 6 BYWASH CHANNEL OVERBRIDGE, BOHERNABREENA RESERVOIRS. This bridge [FIG. 192] replaced in 2006 the original bridge shown in background of FIG. 146. The original bridge was connected to the upper dam and road between the bywash channel and upper reservoir.

NO. 7 Bridge [FIG. 193] giving access to upper dam across the upper spillway, Bohernabreena Reservoirs built 2006.

NO. 8 Existing footbridge below upper spillway [FIG. 147] built circa 1885.

NO. 9 Bridge over lower spillway to lower dam [FIGS. 8 & 186]. This was replaced by a new single span bridge [FIG. 187] over the reconstructed spillway in 2006.

NO. 10 New footbridge below lower spillway [FIG. 191].

NO. 11 FORT BRIDGE. Built after 1821 (around 1837, according to Joyce) originally known as Callaghan's Bridge [FIG. 9] or Bohernabreena Bridge, replaced a ford and footbridge. It carries the original 1886 watermain from Bohernabreena to Ballyboden. In the riverbed below the bridge, there is a deep depression known as the Sheep Hole and into which the water was precipitated over a high weir.

NO. 12 OLDBAWN BRIDGE. The original three-span bridge, built in 1800, was weakened by the frequent floods and replaced by a single-span structure in 1840, [FIG. 10]. Because of the lack of connecting roads in the area, a ford at this location was still in use after the 1840 bridge was built. In a field beside the ford a gruesome triple execution took place in 1816 when the Kearneys—a father and two sons—were hanged for a murder that had both political and agrarian overtones.

In 1879, William Handcock expressed concern for the stability of Oldbawn Bridge, noting that its foundations were exposed.

NO. 13 MOUNT CARMEL PARK. This is a footbridge at the Balrothery Weir. Handcock records that in the early to mid-1800s, it was possible to ford the Dodder downstream of the Balrothery Weir. In time, however, riparian changes eliminated this crossing and for many years the only way of crossing the river at this point during a flood was by wading along the top of the weir. This was an extremely dangerous practice—one slip could have fatal consequences.

A wooden footbridge erected at this location by public subscription was swept away. It was replaced around 1860 by an iron lattice bridge, which in turn was superseded by the present structure, opened in 1995.

NO. 14 M50 MOTORWAY. This is the former Western Parkway. The Dodder flows under the motorway just south of Junction 10. There are four bridges at this location. The motorway northbound and southbound bridges are on common

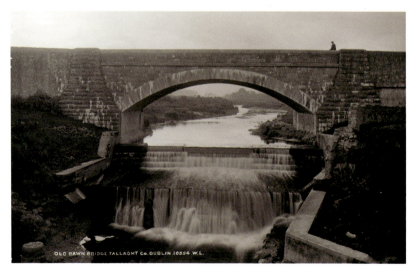

FIG. 8
Bridge to lower dam and reservoir, Rathmines Waterworks, Bohernabreena (Robert French, c. 1900)

FIG. 9
Callaghan's Bridge, (Fort Bridge) Bohernabreena (Robert French, c. 1900)

FIG. 10
Old Bawn Bridge circa 1900 (Robert French, c. 1900)

abutments. Two other bridges—one at each side of the motorway—carry the roads affording entry and exit to and from the roundabout [FIG. 11].

NO. 15 NEW SPAWELL BRIDGE, which connects Wellington Lane with Firhouse Road, was opened in 2001.

NO. 16 TEMPLEOGUE BRIDGE (also known as Old Bridge and Austin Clarke Bridge) was built in 1798 and attributed to a Mr. Bermingham. This individual, Handcock wrote, was "like the noble lord who, of his great bounty, built a bridge at the expense of the county." The danger of Templeogue Bridge being undermined was reduced in the nineteenth century by the construction of a dam or weir downstream.

The original bridge was demolished during road widening operations and a new, wider bridge was built, and opened in 1984. A plaque on the bridge bears the inscription:

Austin Clarke Bridge
Opened by Councillor Mrs. Bernie Malone
Chairman Dublin County Council
11th December 1984
Chief Engineer – Brendan Murphy B.E., C.Eng., F. IEI.

The poet and author Austin Clarke (1896-1974) lived beside the old bridge and his house had to be demolished to make way for the new road and bridge. Only the new bridge was named after Clarke, the older one being generally known as Templeogue Bridge or Old Bridge.

NO. 17 SPRINGFIELD AVENUE. This bridge connects Templeville Road with Dodder Park Road [FIG. 12].

NO. 18 BUSHY PARK. Immediately upstream of where the Owendoher River joins the Dodder, this footbridge gives access from Springfield Avenue to Bushy Park.

Immediately downstream of this bridge there is a ford or series of stepping stones [FIG. 13] which are a feature of the park. This is to remind people that before bridges were built this was the only means of crossing the Dodder.

NO. 19 PEARSE BRIDGE is also known as Big Bridge and Rathfarnham Bridge, [FIG. 14]. Michael Fewer in *The Wicklow Military Road* notes that a number of timber structures spanned the river at this location, the earliest dating from about 1381. There are records which state that a Joan Douce, of St. Audeon's parish in Dublin, contributed the sum of one mark towards its construction. Evidence that this was a busy and important crossing point even long before this, however, was found when, in 1912, during the excavations for a drainage scheme, a nine-foot-wide causeway or ford, built of great slabs of stone, was found crossing the course of the river. The stones bore the parallel groves from many years of passage of wheeled traffic.

FIG. 11
M50 Bridges, Balrothery
(Don McEntee)

FIG. 12
Springfield Road Bridge
(Don McEntee)

FIG. 13
Pedestrian ford, River Dodder,
Bushy Park - (Don McEntee)

FIG. 14
Pearse Bridge, Rathfarnham
– original bridge in stone and
widening in concrete
(Don McEntee)

In 1652, Gerard Boate wrote "to go from Dublin to Rathfarnum (note the spelling), one passeth this river upon a wooden bridge which although it be high and strong, nevertheless hath several times been quite broke and carried away through the violence of sudden floods." A new bridge was built in 1659 and replaced in 1728. This structure, which had an arch a hundred feet wide, was destroyed by a flood in 1754 and rebuilt in 1765. It was widened in 1953. The photograph shows the existing stone arch bridge and the concrete bridge.

NO. 20 DODDER ROAD. This footbridge, at the junction of Lower Dodder Road and Woodside, affords access to Orwell Park.

NO. 21 ORWELL BRIDGE. The single arch bridge [FIG. 15] was built in 1848 by Patrick Waldron of Orwell House. It is also called Waldron's Bridge. In the background can be seen the Mill Race House at the start of the mill race to Dodder Mill No. 22. The 1841 painting by William Howis [FIG. 16] is probably of an earlier wooden bridge at this location. The 1843 OS map shows a wooden bridge at this location. It was replaced in 1970 by a three spans, reinforced concrete bridge [FIG.17].

NO. 22 ORWELL WALK. This footbridge is one of two affording access from Orwell Walk and Gardens to Dartry Park.

NO. 23 DARTRY COTTAGES. This is the second footbridge connecting Orwell Walk with Dartry Park.

NO. 24 CLASSON'S BRIDGE. Francis Ball in his *History of County Dublin* notes that Classon's Bridge [FIG. 18] was built by John Classon, mill owner, in the latter part of the eighteenth century and up to then Old Milltown Bridge was the only way of crossing the river. When Classon's Bridge was constructed there was no connecting road on the south side, suggesting that the bridge was built to give access to the quarries on the south side (right bank) of the Dodder. It has three masonry arches and was widened on both its up and downstream sides in 1928.

FIG. 16
A painting of a wooden
bridge at Orwell by
(William Howis, 1841)

FIG. 17
Present day Orwell Bridge
(Don McEntee)

FIG. 18
Classon's Bridge – original
narrow bridge with widening
on both sides (Don McEntee)

FIG. 19
The Nine Arches, Milltown
(Don McEntee)

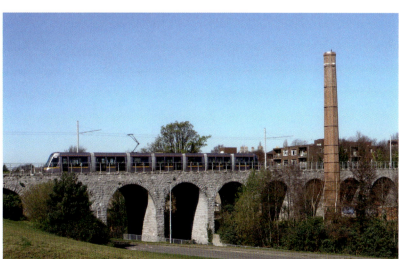

FIG. 20
Old Bridge Milltown
(Packhorse Bridge) circa 1880
looking upstream

FIG. 21
Packhorse Bridge, Milltown
showing river bed excavated
below bridge foundations

FIG. 22
Packhorse Bridge, Milltown
(Don McEntee)

Long before the Dropping Well public house was established, Milltown was home to a thriving silk industry, two corn mills, an iron mill, a paper mill and a sawmill which was located at Classon's Bridge on the site of the Dropping Well. The bridge built by John Classon, when he started his milling venture in the late 1700s aiding transport to and from his business, was of limestone lifted from the bed of the Dodder.

NO. 25 THE NINE ARCHES, MILLTOWN. Built 1852/53 to carry the Harcourt Street-Bray railway line, [FIG. 19]. This bridge was disused from 1959 until 2004 when Luas trams began running over it.

NO. 26 PACKHORSE BRIDGE, MILLTOWN. This partly ruined bridge consists of three spans, two over the river, the other over a mill race to Dodder Mill No. 24. FIG. 20 shows the bridge in a bad state of repair circa 1880. In *Irish Stone Bridges – History and Heritage* by O'Keeffe and Simington there is a detailed description of Milltown Old Bridge also known as Packhorse Bridge. Nobody has as yet been able to establish exactly how old this bridge is and it is probably the oldest surviving bridge in Dublin. Some historians have attributed the bridge to Sir Peter Lewys, who built the bridge of Athlone in 1567 and quarried stone for repairs to Christ Church Cathedral from the Dodder at Milltown in 1565. Packhorse Bridge is shown on the Down Survey map compiled in 1656.

The two abutments of the bridge are founded on rock [FIG. 21]. The central pier is founded on a pedestal of solid rock, suggesting that that the rock in the river bed had been quarried out after the construction of the bridge. The bridge is not in public ownership and, given the nature of sixteenth and seventeenth century activities of milling and quarrying in the area, it was most likely a private bridge.

Packhorse Bridge [FIG. 22] is a frequent choice of subject for artists. Now traversed only as a footbridge, in the eighteenth and nineteenth centuries it was used intensively by excursion traffic to and from Enniskerry. At that time, Milltown was a staging point on the journey, where refreshments were also served.

NO. 27 MILLTOWN BRIDGE. Samuel Brocas' sketch [FIG. 23], circa 1820, downstream of the bridge, shows the single arch bridge with the River Slang mills in the background. FIG. 24, circa 1880, shows the bridge with a house on left bank. In the photograph the river bed is exposed with very little flow in the River Dodder. This was before the reservoir for Millers' compensation water was constructed in Bohernabreena. The bridge was widened with a new span upstream of the old one.

NO. 28 CLONSKEAGH BRIDGE. The present single span bridge was widened with a concrete section on the upstream side in 1952. In "Down the Dodder" (Wolfhound Press, 1991), Christopher Moriarty notes that this bridge is 18 metres above sea level and 3.7km from the South Wall.

A painting by Thomas Roberts in 1770 [FIG. 94] shows a two span arched masonry bridge at Clonskeagh.

FIG. 23 (LEFT)
Milltown Bridge Co. Dublin
(sketch by Samuel Frederick
Brocas, circa 1812 –1847)

FIG. 24 (BELOW)
New Bridge Milltown
(Charles Russel, circa 1880)

FIG. 25 (TOP)
Old Bridge Clonskeagh -
Allens (circa 1800 - 1820)

FIG. 26 (BELOW)
Metal footbridge , Beaver Row
(Don McEntee)

A painting by Allens in the National Library shows a two span arched masonry bridge at Clonskeagh, the period quoted being 1800-1820 [FIG. 25].

A painting by J Cunningham, circa 1920 [FIG. 92] shows a single arch bridge across the river.

NO. 29 BEAVER ROW. A wooden footbridge, connecting Beaver Row with Brookvale Road, was constructed circa 1811 by the Wright Brothers who owned the Beaver Hat Factory. This was to facilitate the workers living on Beaver Row access to the hat factory on the north side of the River Dodder. The wooden bridge was replaced with an iron footbridge with mid-span support between 1876 and 1880 [FIG. 26]. The span is 20 metres across. Wrought iron was used for the span structure and its latticed railings and cast iron for the support column. The bridge was refurbished and strengthened in 2015.

NO. 30 DONNYBROOK (ANGLESEY) BRIDGE. The first bridge, four spans, at this location [FIG. 27], which replaced a ford in the seventeenth century, was destroyed in a storm on 11th September 1739. As can be seen in Francis Place's painting the river between the banks was very wide and shallow. Lord Fitzwilliam's agent wrote on 14th June 1740 that "The bridge at Donnybrook that was demolished last winter, we have got a Presentment to rebuild, and if Lord Allen will be, in proportion, as generous, as Baron Wainewright, the same will be accomplished this season".

The new bridge, built in 1741, was destroyed six months later. Samuel Sproule's drawing of 1780 shows a three span bridge at this location [FIG. 28].

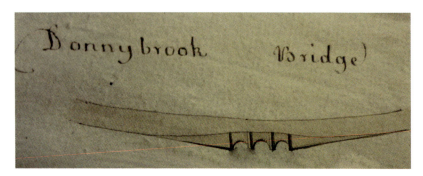

FIG. 27
Dublin from Donnybrook
Bridge with four span
bridge at Donnybrook
(Francis Place, 1698)

FIG. 28
Three span Donnybrook Bridge
(Samuel Sproule, 1780)

FIG. 29
Single span Anglesey Bridge,
Donnybrook (Don McEntee)

The present Anglesey Bridge, a single span arch bridge, [FIG. 29] was erected in 1832 and named for Henry Richard Paget, Marquis of Anglesey, and Lord Lieutenant. At the time this bridge was built the river channel has been reduced in width and canalised with a stone wall on right bank and mixture of stone wall/rock outcrop on left bank.

NO. 31 ANGLESEA ROAD. This bridge, opposite the RDS, was built in the 1990s to give access to apartments built on the site of the former Johnston Mooney & O'Brien bakery.

NO. 32 BALLSBRIDGE was formerly referred to as Ball's Bridge and erected as a four span bridge in 1751 [FIG. 30]. It was rebuilt in 1791 and again in 1835 as a three-span masonry structure (as illustrated in Arthur O'Connor's painting of 1830 [FIG. 99]. Trams are shown passing over the narrow bridge in FIG. 31. It was rebuilt and widened in 1904 [FIG. 32] at the expense of the Dublin United Tramways Company.

Samuel Ossory Fitzpatrick's 1907 volume *Dublin – a Historical and Topographical Account of the City*, re-issued in 1977, refers to the Ballsbridge area. The drowning of Alderman John Usher, who in 1629 tried to cross a ford on the Dodder when it was in spate (in flood), is recounted. Fitzpatrick then goes on to suggest that the Bridge of Symon's Court (Simmonscourt) was built near the site in 1637, possibly where Ballsbridge now stands.

Downstream of Ballsbridge, at Beatty's Avenue, a main sewer (already described) formerly crossed the river on piers, which were an impediment to branches and other large debris carried down in times of spate. The sewer was relaid as a siphon under the river in 2000.

NO. 33 LANSDOWNE ROAD RAILWAY BRIDGE. In November 1834, a month before the Dublin & Kingstown Railway was due to open, there was a serious flood in the Dodder. Some uprooted trees carried away the timber being used for the rebuilding of Ball's Bridge. The wreckage jammed under the new railway bridge, which was destroyed when the timber rose with the water. A temporary replacement which was erected in ten days was rebuilt in 1847 and lasted until 1851 when a new iron bridge was built. During the 1931 flood, this bridge fell into the river and was replaced by a steel girder structure.

A major rainstorm on 24th October 2011 brought considerable quantities of debris down the Dodder. Much of this lodged under the railway bridge in a manner similar to that of December 1834. The bridge was inspected and closed for four days while repair work was undertaken.

NO. 34 FOOTBRIDGE BESIDE RAILWAY BRIDGE.

NO. 35 NEW BRIDGE, Lansdowne Road, also known as Herberts Bridge. The present day bridge is a masonry three spans bridge, FIG. 33, built in the mid eighteenth century.

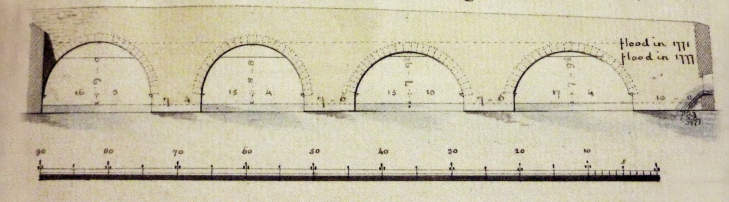

Balls-bridge

N.3 in the year 1771, A flood filled the arches of this Bridge & overflow'd the road

in the year 1777 A flood arose within one foot of the top of arches

For which body of Water, there <u>must</u> be provision { Area of flood in 1771,. 507 feet
{ Area of flood in 1777. 437.1

Flood in 1771
Flood in 1777

Elevation of Balls-bridge, by a scale of 15 to an Inch

FIG. 30
Balls-bridge
(Samuel Sproule, 1780)

FIG. 31
Bridge at Ballsbridge
circa 1900 before bridge
was widened

FIG. 32
Bridge at Ballsbridge
(Don McEntee)

FIG. 33
New Bridge, Lansdowne Road
(Don McEntee)

FIG. 34
Wooden bridge over Dodder
on road from Beggarsbush to
Irishtown (James Cullen, 1693)

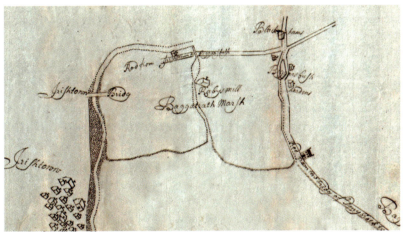

J Cullen's map of 1693, FIG. 34 shows a road from Beggarsbush to Irishtown with a wooden bridge crossing the Dodder. This bridge was located in the vicinity of the present day New Bridge. The wooden bridge was no longer used when a stone bridge was built in Ringsend in 1729.

NO. 36 LONDONBRIDGE ROAD. The first bridge here was a wooden structure, of which the National Gallery has an 1841 coloured drawing by GV Du Noyer [FIG. 35]. Named Londonbridge by Campbell in 1811, this bridge was washed away in a storm and replaced by a three-span arched masonry structure in 1856-57 (de Courcy). A short distance downstream of Londonbridge, trunk sewers of the Rathmines and Pembroke main drainage scheme, already described, siphon under the river to the site of the former Londonbridge Road pumping station.

NO. 37 RINGSEND BRIDGE. The Dodder originally entered the sea through a number of channels in the form of a delta. As a result, access to Ringsend was by a bridge at Ballsbridge, by crossing the river at low tide, or, later, by hiring what was known as a "Ringsend Car". Petitions were made in the early 1600s

for a bridge over the Dodder at Ringsend, but the first bridge at this location was not built until 1650. It is referred to by Gerard Boate in his 1652 *Natural History of Ireland*. Its exact location has been questioned by Professor John de Courcy in *The Liffey in Dublin*, who suggests that it may have been nearer to Ballsbridge. Weston St. John Joyce's *The Neighbourhood of Dublin* states that the stone bridge referred to by Boate was where Ballsbridge now stands. In Thomas Phillips' MAP NO. 8 (1685) a stone bridge is shown at Ballsbridge.

Accounts survive, probably accurate in the topography of the period, of long gone bridges at locations between Ballsbridge and Ringsend from the 1600s onwards. Because so much of what is now Ringsend was undeveloped and tidal for centuries, these sites cannot be accurately pinpointed today. Gerald Boate refers, around 1650, to a bridge over the Dodder on completion of which the river changed course, bypassing the structure.

In Bernard de Gomme's map of 1673 a bridge is shown in a marsh area where the Dodder bifurcated. On Cullen's map of 1692 [MAP NO. 9], there is a wooden bridge over the Dodder approximately at Lansdowne Road. Cullen's map of 1706 [MAP NO. 10] shows a wooden bridge at Irishtown.

FIG. 35
Old London Bridge, on the River Dodder, Dublin (wooden toll bridge)
(George Victor du Noyer, 1841)

A replacement bridge, which was erected in 1727, is shown on E Cullen's map of 1731 [MAP NO. 11] and was the subject of correspondence to Lord Fitzwilliam from his servant in Ireland. On 20th March 1739 this official wrote that he was waiting for "Presentment money charged in my last account for repairs to the Bridge at Ringsend". On 14th June 1740 he wrote that "The money I advanced for Ringsend Bridge, will, in its course of Presentment Payments, come to my hand and which I will remit when paid to me"

Samuel Sproule's map of 1780 shows Ringsend Bridge as having four spans [FIG. 36]. It was destroyed by a flood in 1782. A 1786 partially constructed replacement of questionable stability was washed away in 1787. It gave way to a new structure, built in 1789 at a cost of £815, which in turn was destroyed by the notorious storm that also wrecked the Ormonde Bridge over the Liffey in January 1802. Francis Ball's book an illustration by John James Barralet, FIG. 37, shows the remains of Ringsend Bridge after its inundation.

The presence of a bridge at Ringsend was considered important to maintain access to the markets for fish deliveries, passengers and mail from ships and the barracks at Poolbeg. A temporary toll bridge was built in 1807 with the permissions of the Grand Canal Company and in 1808 the £8,000 funds which had been levied on the citizens were stolen by the treasurer, a Mr. Baker

The natural obstructions to building a bridge at this place was the location of the actual river channel, the very high flows resulting from heavy rainfall, storms in the upper region of the river and the difficulty with locating good foundations in a tidal estuary which was also indirectly impacted on by its close proximity to the River Liffey.

John Semple (1763-1841), who was appointed as architect for the new Ringsend Bridge in 1810, was reported to the Grand Jury for being slow in prosecuting the work. He responded that very bad weather, with heavy rain and high winds, had held the work back, also pointing out that he had the materials on site. Semple was very experienced in design and construction and was probably the son of George Semple, who had built Grattan (Capel Street) Bridge in 1755 and had written a celebrated "A Treatise on Building in Water".

It is possible that Semple was both architect and contractor for Ringsend Bridge as he is recorded as having sold off timber and scaffolding after completion of the bridge in 1812 [FIG. 38]. The single span is part of a complete ellipse built on the inverted arch principle used by William Jessop in the construction of the Grand Canal Basin. The great strength of this form is that the loads imposed on the arch by traffic are not solely resisted by the abutments, as is normal, but transferred to the full width of the river bed by the inverted arch. The stone forming the arch is chamfered similar to the French "cornes de vache" and the abutments are curved so that the oval form is continued on the bed of the river. In this way, while the soil in the river bed is a weak soil, it was used to its best advantage by spreading the load of the bridge over the same area but underneath it. This type of structure also allowed for settlement to occur as the soil is compressed by acting as single unit so that differential settlement was minimised. These elements enhance and protect the bridge by improving the flow of the river through the bridge and reducing the risk of scouring to the

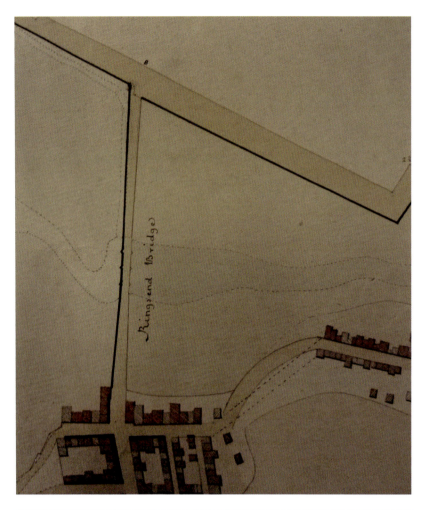

FIG. 36
Four span Ringsend bridge
(Samuel Sproule, 1780)

FIG. 37
Ringsend bridge after
inundation (John James
Barralet, 1802)

FIG. 38

Ringsend Bridge 1812 (showing elliptical arch and part of elliptical base) (Don McEntee)

foundations. This remarkable bridge, which was originally named for Queen Charlotte, has endured for more than 200 years.

NO. 38 NEW BRIDGE ACROSS THE GUT. Professor John de Courcy describes a proposal, in 1812, for a bridge across the mouth of the Dodder, to connect Sir John Rogerson's Quay with the South Wall. It reached the stage of a tender being accepted, but the project was then abandoned.

In 2015, Dublin City Council ordered design work to begin on a new bridge at this location to be completed by 2020.

There are further references to bridges in Ringsend in the section of Chapter 4 describing villages on the Dodder.

WARTIME TANK TRAPS ON BRIDGES

In the early days of World War Two or "The Emergency" as it was known in Ireland, when there was a credible fear of invasion, tank traps were erected on some bridges spanning the Rivers Tolka and Dodder. Rathfarnham Bridge on the Dodder [FIG. 14] was one such. The tank traps consisted of two very substantial concrete and steel barriers across the roadway at either end of the bridge. Two open sections, wide enough to allow a bus to pass and capable of being closed rapidly with steel rails, were provided in each barrier. The openings were staggered chicane fashion, demanding a high degree of skill and attention from drivers. The tank traps, which also had protruding spikes facing oncoming traffic and constituting a serious danger for the unwary, were removed quickly at the end of the war.

CHAPTER 8

ꟻHISTORY OF MILLS IN THE DODDER CATCHMENT

THE MALLET REPORT — INDUSTRY ON THE DODDER

The Commissioners for Drainage and the Improvement of Water-power, appointed under an Act of 1842, engaged Robert Mallet to examine the Dodder. His brief was to make recommendations for the creation of reservoirs on the river "for the prevention of sudden floods, and accumulation of water for the constant supply of mills on that river and on the watercourse." The watercourse referred to was the Poddle.

During periods of low rainfall or prolonged drought the flow in the Dodder fell to a trickle and the mills had to cease production. FIG. 24 shows the dry river bed at Milltown circa 1880.

Robert Mallet's magisterial report, dated 17th January 1844 and published by Her Majesty's Stationery Office (HMSO), described the effects of serious flooding: "A few years only have elapsed since one of the frequent floods of the river Dodder occurred so suddenly, at night, and attended with such appalling circumstances, of destruction to property and danger to human life, as to have aroused the public concern for the damage resulting; and which finally led to expensive litigation betwixt parties subjected thereto, and to the subsequent compulsory expenditure of large sums in the execution of works near the mouth of the river, only intended to ward off the destructive effects of future inundations, but not attempting to control them. Yet similar floods, though fortunately not to such a formidable extent, occur upon this river many times every year."

Apart from ensuring constant water supplies to the millers, other benefits would have resulted from Mallet's proposed dams. Around 110 acres of good land, liable to erosion and flooding, would have been made available, and another 70 acres of barren shingle, rendered productive of valuable crops.

An industry that tends to have been overlooked in times past was quarrying, several sites being in use at different periods along or adjacent to the Dodder. Quarrying relies on water to process what is produced, the Dodder and its tributaries being the source used by several quarries. Quarrying relies on water in the production of stone.

MILLS: THE BEGINNING

To fully appreciate the historical significance of the mills along the Dodder, a brief look at how important water has been in the landscape of Ireland over the years is informative.

The earliest settlers came to Ireland by sea and settled along the coast and the navigable sections of rivers. As the forests were removed and the population spread through the country, the rivers and streams became a vitally important part of the locality. Initially they provided clean water for domestic and agricultural uses. As communities developed the potential of rivers and streams as a source of power was recognised.

For thousands of years corn has been grown as a food source. The early inhabitants of Europe and Asia ground the grain using rudimentary hand tools (known as quern stones) which separated the edible kernels from the chaff. Quern stones were generally replaced by millstones once mechanised forms of milling appeared, particularly the water mill and the wind mill, although animals were also used to operate the millstones. However, in many non-Westernised, non-mechanised cultures they are still manufactured and used regularly and have only been replaced in many parts of the world in the last century or so.

In China there are records showing horizontal waterwheels in use in 200 BC; these were applied to rotary millstones.

In the second century AD vertical undershot water mills were in use in Europe. The use of water mills spread through Europe in the fourth and fifth centuries and were used mainly in the grinding industry. The Greeks invented the two main components of watermills, the vertical waterwheel and toothed gearing, and were, along with the Romans, the first to operate an undershot water wheel, FIG. 39, overshot water wheel, FIG. 40, and breastshot water wheel, FIG. 41.

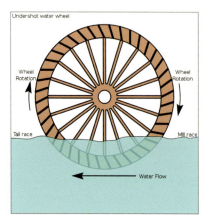

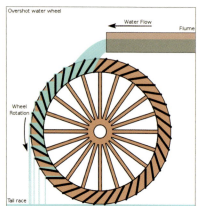

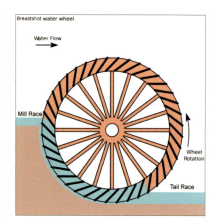

The ancient world gave birth to the vertical waterwheel and nurtured the early stages of its growth. It was medieval Europe that brought the vertical waterwheel through adolescence and into adulthood.

The horizontal water wheel is the simplest form of water wheel, which is immersed in the flow of the river. The flow of the water turns the wheel which transfers the power vertically to rotary millstones overhead. The configuration for horizontal watermill is shown in FIG. 42 There was no gearing involved. This is the most inefficient form of water power. Colin Rynne's sketch [FIG. 43] shows a ninth century monastic horizontal mill on High Island, Co. Galway.

The vertical waterwheel is much more efficient than the horizontal waterwheel and delivers more power. The undershot wheel is simply set into the flow of the river, or in a millrace, with the main thrust of the flow driving the wheel as a simple paddle. The configuration for undershot waterwheels is shown in FIG. 44. The flow to the undershot water wheel in FIG. 45 is controlled by raising or lowering the sluice gate in the channel beside the water wheel.

The overshot wheel is fed with the water flow to the top of the wheel. The water fills buckets built into the wheel and the weight of the water starts to turn the wheel as shown in FIG. 46. The water spills out of the bucket on the downside into a tailrace leading back to the river.

Overshot wheels require the construction of a dam on the river above the mill and a more elaborate millpond, sluice gate, mill race and spillway or tailrace. The configuration for overshot waterwheel is shown in FIG. 47. The overshot wheel is around two and a half times more efficient than the undershot. The breastshot wheel is similar to the overshot, with the water directed to the middle of the wheel. The breastshot wheel is more efficient than the undershot but not as efficient as the overshot.

The vertical waterwheel produced rotary motion around a horizontal axis, which was used, with cams, to lift hammers in a forge, fulling stocks in a fulling mill and other types of machinery. However, in corn mills rotation about a vertical axis was required to drive its millstones. The horizontal rotation was converted into the vertical rotation by means of gearing, which also enabled the millstones to turn faster than the waterwheel [FIG. 48]. Over the centuries the operational systems inside the mills were refined but the basic principle of water power remained the same.

Of all the inanimate sources of power (not using man or animal power) developed in the Middle Ages the water wheel was, by far, the most important. In the centuries following the demise of the Roman Empire the water wheel spread from a few small pockets to practically every region of the European continent. In the early centuries the watermill had usually been tied to aqueducts. In the medieval period water wheels were put into operation on streams of every size, from small mountain brooks to substantial rivers; they were even put to work on tidal inlets. In the earlier times water wheels had been used primarily for flour milling. By 1500 they were applied to a host of different industrial processes, some on a very substantial scale. European society enthusiastically adopted water power and incorporated it into the predominant feudal-manorial system.

FIG. 39
Undershot water wheel

FIG. 40
Overshot water wheel

FIG. 41
Breastshot water wheel

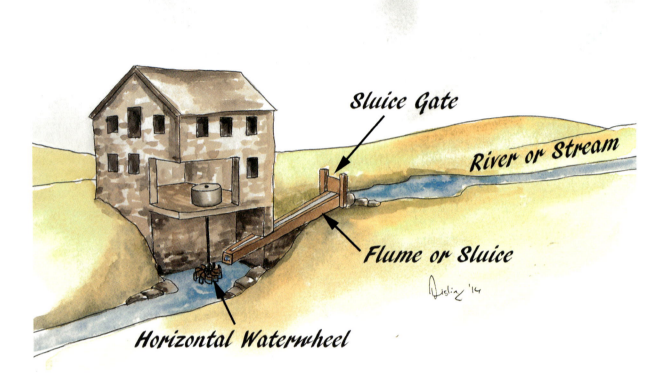

Sluice Gate

River or Stream

Flume or Sluice

Horizontal Waterwheel

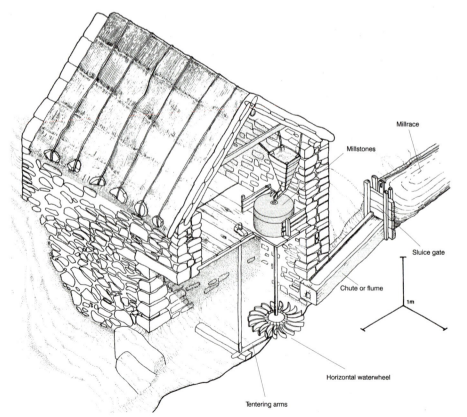

Millrace

Millstones

Sluice gate

Chute or flume

1m

Horizontal waterwheel

Tentering arms

FIG. 42
Sketch of horizontal
waterwheel
(Aisling Walsh and
Don McEntee)

FIG. 43
Reconstruction of ninth
century monastic horizontal
mill on High Island,
Co. Galway (Colin Rynne)

River or Mill Race

Undershot Waterwheel

FIG. 44
Sketch of undershot
waterwheel
(Aisling Walsh and
Don McEntee)

FIG. 45
Undershot waterwheel
at Braine-le-Château on
Hain River, Belgium with
sluicegate controlling flow
to waterwheel

FIG. 46
Overshot vertical waterwheel,
Crannford Mill (Don McEntee)

DRIVE SHAFT

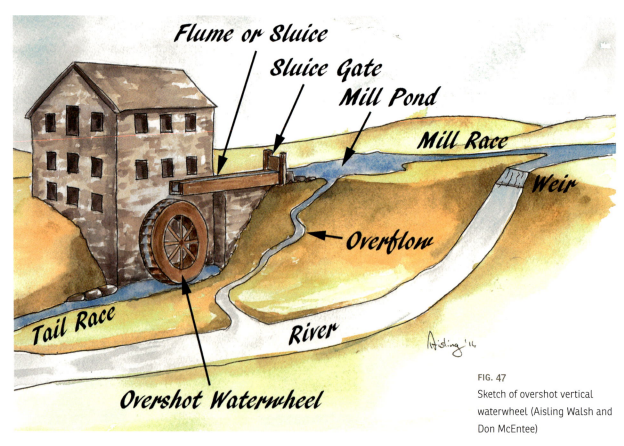

Flume or Sluice

Sluice Gate

Mill Pond

Mill Race

Weir

Overflow

Tail Race

River

Overshot Waterwheel

FIG. 47
Sketch of overshot vertical
waterwheel (Aisling Walsh and
Don McEntee)

FIG. 48

Corn mill gearing and
power transmission
(Colm McEntee)

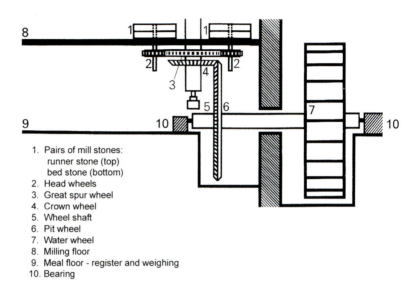

Corn Mill Gearing and Power Transmission

1. Pairs of mill stones:
   runner stone (top)
   bed stone (bottom)
2. Head wheels
3. Great spur wheel
4. Crown wheel
5. Wheel shaft
6. Pit wheel
7. Water wheel
8. Milling floor
9. Meal floor - register and weighing
10. Bearing

MILLS IN IRELAND

The milling of corn using millstones in Ireland has its origins in the first millennium.

PW Joyce (*The Origin of Irish Names of Places,* 1871) refers to mills and kilns:

"Many authorities concur in showing that water mills were known in Ireland in very remote ages, and that they were even more common in ancient than in modern times. We know from the lives of the Irish saints, that several of them erected mills where they settled shortly after the introduction of Christianity, as St. Senanus, St. Ciaran, St. Fechin. In some cases mills still exist on the very sites selected by the original founders. An example is at Fore in Westmeath where 'St. Fechin's mill' works as busily today as it did twelve hundred years ago.

"It appears certain that water mills were used in Ireland before the introduction of Christianity. The first reference to water power in Ireland is attributed to King Cormac MacArt. In the third century the annals record that Cormac sent across the sea for a millwright which the King of Scotland supplied. He constructed a mill on the stream of Nith, which flowed from the well of Neamhuach at Tara. This was the first recorded mill ever erected in Ireland and a mill still occupies its site."

As pointed out by Hugh Fogarty, P.W. Joyce was writing in the nineteenth and early twentieth centuries, before the earliest records of Irish were adequately understood, and Joyce's work reflects that. The historicity of many of his references is no longer regarded as being entirely reliable.

Although various Irish 'annals' mention Cormac mac Airt, those 'sources' are thoroughly untrustworthy from a historical point of view, and the actual existence of Cormac mac Airt himself is by no means certain — what is certain is that every 'annal' or 'historical record' that purports to recount something that he did is without doubt a more recent invention, and in some cases no older than the seventeenth century.

Alongside evidence from archaeological investigations (most importantly, the work of Colin Rynne) there is a significant amount of documentary evidence attesting to the existence and use of water mills in the early history of Ireland, and the majority of this is in early Irish legal texts. The standard introductory works are those by Fergus Kelly: (Dublin, 1988, frequently reprinted) and *A Guide to Early Irish Law* and *Irish Farming* (Dublin 1997).

The most important single law-text regarding mills is the — probably seventh-century — *Coibnes Uisce Thairidne* (literally 'on the kinship of conducted water'), which is described in Kelly's *Guide to Early Irish Law* "which gives the rules for conducting water across neighbours' land to power a mill". (The text has been edited and translated by Daniel Binchy in volume 17 of the journal *Ériu.)*

Quoting from the journal *Ériu* "*Uisce thairidne* in the present tract it has the specialized meaning of 'mill-race'. Obviously, then... the introduction of water-mills into Ireland is the prerequisite for their earliest appearance in legal sources. Water-mills figure among the many innovations with which Middle-Irish writers have credited Cormac mac Airt, but the first (genuine) reference to them in the Annals occurs about the middle of the seventh century. In the lives of saints who flourished in this and the preceding century water-mills are mentioned on several occasions. Indeed, it may well be that the replacement of querns by water-mills first spread from the monasteries.

Mills could be owned jointly or severally. If water to work the mill has to be conducted across the land of a neighbour, the owner had to pay him a fee which varied according to the nature of the land breached (arable land/ rough land etc.). This was apparently a single payment, not a periodical rent. It should be noted that a neighbouring land-holder was not free to refuse a passage to the water across his land. Further, when the mill-race has been accepted (either by formal agreement or tacit acquiescence) by two successive owners, the mill-owner acquired an absolute title to it as against their heirs. An neighbour across whose land the mill-race flowed could choose a day's free use of the mill at regular intervals... perhaps (once) a year. All other persons could only grind their corn at the mill with the consent on the owner, who normally demanded a certain proportion of the grain as payment.

This is the type of mill we find attached to monasteries and to the estates of the wealthier members of the 'noble grades'. When a mill was erected with the co-operation of the free-holders between the 'source' and the 'pond' each of these free-holders became a partner in the mill, and his 'share' consisted in the right to the exclusive use of the mill for a certain proportion of the period of rotation between the various owners."

FIG. 49
Ship mill with
undershot waterwheel

MILLS IN EUROPE

One of the most striking indications of the spread of power revolution in medieval Europe is provided by the Domesday Book (1080-86), the result of a census ordered by William the Conqueror shortly after the Norman conquest of England. Most small streams in southern and eastern England were covered with mills. In some areas they were placed less than a mile (1.6km) apart. In England there were 5,624 watermills, an average of one for every 50 households. Records suggest that at this time the density of watermills in Ireland was similar to that in England.

By the fifteenth century waterwheels in Europe were powering corn mills, paper mills, tanning mills, fulling mills, cloth mills, saw mills, hammer mills (for iron and copper wares), pounding mills (for extracting oil from nuts and herbs) and mills for polishing arms and sharpening various implements.

The boat mill [FIG. 49] was the earliest modification of the conventional undershot vertical wheel to enable medieval Europeans to tap the power of large rivers with some degree of economy and reliability. A boat mill had no trouble adapting itself to changes in water level or crowded bank conditions. The first record of the boat mill comes from the Gothic siege of Rome in 536 AD. In Ireland, the horizontal wheel, not the vertical wheel, predominated until the eighteenth century.

Before the Industrial Revolution the main sources of power in Europe were wind and water. Windmills were most common in low lying flat areas such as Holland. There was a small number of windmills in Ireland; William Hogg in *The Millers and the Mills of Ireland about 1850* notes that thirteen windmills were recorded in the 1800 directory of County Dublin.

On the banks of rivers and streams water wheels were used to provide power for diverse industries. Every suitable watercourse was harnessed for its energy and Hogg estimates that there were more than 7,000 mills spread across Ireland when milling was in its heyday.

TYPES OF MILLS

In *Art and Architecture in Ireland,* Volume IV, Livia Hurley wrote: "The term 'mill' encompasses a range of industrial structures including flourmills, textile mills, paper mills, spade mills, powder mills and sawmills, all of which prior to the advent of electricity, were powered by water from a river...." Most of these industries were represented by the mills on the Dodder. Factories, often known as manufactories in the past, were sometimes also called mills.

WA McCutcheon in *The Industrial Archaeology of Northern Ireland* describes in detail the evolution of the milling industry in the north of Ireland and the mechanics of how mills worked. Due to the demand for power to run the expanding industries in the eighteenth and nineteenth centuries, and especially the linen industry, great strides were made in developing the milling technology.

The most common types of mill were:

FIG. 50 (LEFT)
Hand trough quern

FIG. 51 (ABOVE)
Saddle Quern and Rubbing
Stone (H. Claire)

CORN MILLS. The most ancient grinding machine of all, and most difficult and laborious to work, was the quern-stone. The earliest forms of quern were the saddle and trough querns. The earliest quern so far discovered dates to c. 9500–9000 BC and was found at Abu Hureyra, Syria. A later development was the rotary quern.

In a trough quern the grain was placed on in the centre and ground with a hand held stone known as a grain rubber. FIG. 50 shows a trough quern.

The saddle quern [FIG. 51] consists of two parts: the lower stone (bedstone) which is shaped like a saddle and the upper handstone (rubber). The handstones for saddle querns are generally roughly cylindrical (not unlike a rolling pin) and used with both hands or rough hemispheres and used with one hand.. This provides a crushing motion, not a grinding action and is more suitable for crushing malted grain. The constant grinding motion of using a quern was very tiring work. It was not easy to produce flour from a saddle quern with unmalted grain.

Saddle querns were gradually replaced by rotary querns which were easier to use and caused less physical strain. A rotary quern consists of two disc shaped stones with holes through the middle and on the outer edge. A handle was inserted to turn the upper stone. The base was stationary and the top rotated by hand [FIG. 52]. The grain was fed through the hole in the middle of the upper stone. The rotary quern used circular motions to grind the material. The handstone of a rotary quern was much heavier than that of saddle quern and provided the necessary weight for the grinding of unmalted grain into flour. The grinding surfaces of the stones fit into each other, the upper stone being slightly concave and the lower one convex.

This type of milling was used extensively in pre and early Christian times. The use of the quern in medieval Ireland was always associated with small-scale domestic production, as it continues in many Asian countries to this day.

The early water mills were little more than an extension of the quern mill, but with a larger pair of grinding stones. The water wheel, not much larger than a cart wheel, was positioned flat on a projecting axle on the bottom of a stream. The water came through a stone or a timber spout to drive the paddles

FIG. 52
Irish rotary quern

FIG. 53
Grain hoist raising bag
of grain to grain loft,
Crannford Mill

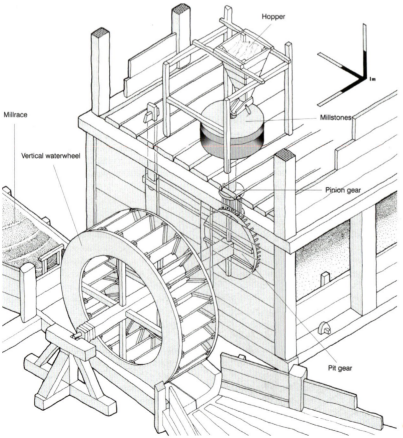

FIG. 54
Reconstruction of undershot
vertical waterwheel mill at
Little Island, Co. cork, circa
630 AD. This is, on present
evidence, the earliest known
example of its type in Ireland
(Colin Rynne)

FIG. 55

Mill pond and midshot
vertical waterwheel, Anamoe
Mill, Co. Wicklow

which turned the main vertical shaft, revolving the top millstone; this system also operated the sieving process.

The undershot [FIG. 39] and overshot [FIG. 40] types of waterwheel became very popular in Ireland in the seventeenth century. Patented systems were developed to use the power from the water wheel to operate elevators, conveyors and sieves. The patent technology came to this country from Scotland. Grain was carried up to the top floor using a power-driven hoist [FIG. 53] and was fed by gravity to grinding stones, sieves and screens producing flour, meal etc depending on the requirements of the farmer. The oldest mill discovered to date in Ireland, with undershot vertical waterwheel, was excavated in Cork by Colin Rynne and dates back to 630 AD [FIG. 54].

Every district in Ireland had a corn mill. A good example is the Anamoe Mill, Co. Wicklow using a midshot vertical waterwheel [FIG. 55]. In the middle ages these were owned and operated by the local landlords. The lord's tenants were legally bound to bring all of their cereals to the lord's mill to be ground. When the farmer got his grain milled he generally paid for this service by giving a percentage of the grain to the landlord and the miller. In this way the landlord used the mill to extract rent indirectly from his tenants. In the eighteenth and nineteenth centuries the native Irish grew corn which was ground into oaten meal for family use and as payment of rent. Most corn mills had drying kilns attached, which were used to dry the grain before milling.

Ownership of water-mills or rotary querns by tenants was outlawed in rural Ireland, the lord's agent being empowered to seize and destroy them.

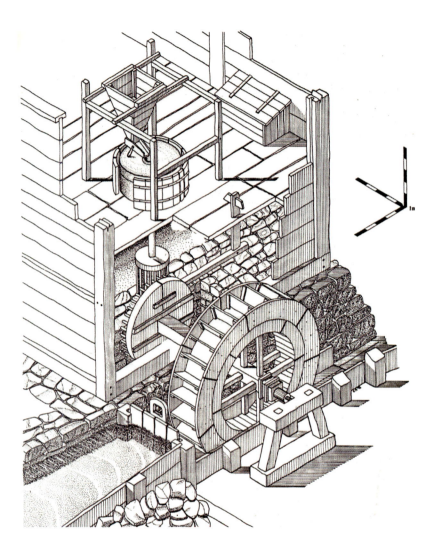

FIG. 56
Reconstruction of fourteenth
century undershot vertical
waterwheel mill at Patrick
Street, Dublin (Colin Rynne)

In Colin Rynne's essay *on The Development of Milling Technology in Ireland c.600-1875* published in Irish Flour Milling, he noted that Ireland in the early medieval period showed a remarkable precocity with regard to the development of water-powered grain mills, a phenomenon that had no parallel anywhere in medieval Europe. There are, indeed, more scientifically dated mill sites of the first millennium in the province of Munster alone than there are from the rest of Europe. The two varieties of water-powered grain mills in use in Ireland from the early decades of the seventh century were the horizontal and vertical undershot wheeled mills. These include the earliest known tide mills.

The vast majority of the water mills erected during the Anglo-Norman period in Ireland were involved in the processing of cereal crops. In 1990 the remains of an Anglo-Norman vertical grain mill were excavated at Patrick Street in Dublin by Colin Wynne. The mill had been built in the thirteenth century and almost rebuilt in the fourteenth century. Its waterwheel was positioned directly over the canalised section of the river Poddle [FIG. 56].

FIG. 57

Diagrammatic ground floor
plan of Crannford Corn Mill,
County Wexford (Colm McEntee)

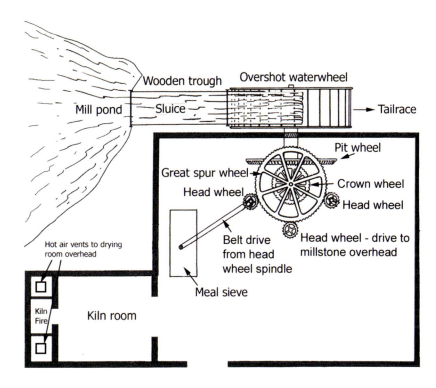

In 1804 the Corn Laws were enacted in Britain and Ireland to protect domestic farmers against foreign competition by the imposition of heavy import duties. After the repeal of the Corn Laws in 1846 hard wheat was imported from the USA and Canada. The hard wheat facilitated the production of white flour and the baking of white bread. With the use of steam power and the introduction of roller mills (from the USA and Hungary) five and six story commercial mills were built in the main port areas of Ireland and inland towns on navigable rivers and canals. In the large cities and towns large scale bakeries began production. This coincided with a huge increase in demand for white bread. The small scale water-mill continued to operate in rural Ireland producing the coarse flour which was used in home baking. In 1890 it was estimated that eighty per cent of the flour sold in Ireland was of the coarse variety.

Michael Lyons started working in 1931 (at six years of age) in the family mill at Craanford in County Wexford and subsequently ran it until it closed in 1948. He describes in detail the working of the mill and its importance in the local community. The ground floor plan of Craanford Mill in illustrated in FIG. 57. The mill gearing and power transmission for this mill is shown in FIG. 48. The internal working of the mill is shown in FIG. 58. Only one of the millstone could be used at any one time. It took about half a day to change the operation from one head wheel to another and to adjust the space between the top runner stone and the bed stone [FIG. 48] depending on the grain to be milled. Upstream of the mill a weir was placed across the River Lask and part of the flow was diverted into a millrace which fed a mill pond [MAP NO. 25]. To operate the mill, water was fed from the pond along a wooden trough to the millwheel. When the wheel rotated power was provided to turn the millstones. The Cranford Mill had three millstones—one to grind wheat to produce flour, one to grind cereal for animal feed and one for crushing oats.

FIG. 59 shows the bed of the millstone for crushing oats without its enclosures. FIG. 60 shows the millstone for grinding wheat with a section of the enclosure. FIG. 61 shows the millstone for grinding animal feed (which was a mixture of oats and barley and sometimes added wheat). The grain was fed from the grain loft above and the directed into the centre hole of the top runner stone and was ground between the two mill stones. The enclosure gathered the ground cereal which was fed to a bag underneath in the case of animal feed and oats. Under the millstone for grinding wheat the ground cereal was fed to a meal sieve at ground floor level. The vibrating sieve sorted the ground flour into one bag and the chaff into the other bag [FIG. 62]. The first millstone [FIG. 59] was made from Cavan Volcanic Rock and was used to crush rather than grind the oats for making porridge. The second millstone used for producing flour [FIG. 60], which accounted for ten per cent of the production, was made from French Burh. This was a very hard wearing millstone which was used in mills throughout the world. The third millstone was made from English Stone Granite [FIG. 61] and was used to grind meal for animal feed—this accounted for eighty per cent of the production.

The local Craanford Mill worked twenty four hours a day and could handle half a ton of grain an hour. With the onset of rural electrification in 1948 the mill closed down as a small local electrically driven modern mill could handle six tons of grain an hour. In rural Ireland all the local water driven mills closed down with the onset of rural electrification.

The drying kiln was used to prepare the corn for grinding by conditioning the corn to get it into a state in which the maximum amount of flour could be extracted at the grinding process. The heat from the fire in the kiln room rose to the floor overhead and was distributed under tiles to produce an overall gentle heat. The corn was spread over the tiles to dry. The skilled miller knew by the touch and feel of the grain how many hours it should be left in the kiln. When the grain came into the mill the miller put his hand into the sack and the distance his arm went into the grain told him how many hours the grain should be in the drying room. There is archaeological evidence that corn-drying kilns were in use in the Bronze Age in Ireland. The earliest recorded dates back to circa 1500BC.

The local corn mill, the school and the church played a pivotal role in the daily life of a village. As Michael Lyons states "the mill to feed you, the school to educate you and to know right from wrong, and the church a place to give

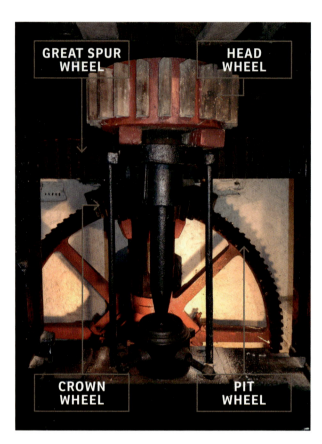

FIG. 58
Transmission gearing from drive shaft to head wheel, Crannford Mill (Don McEntee)

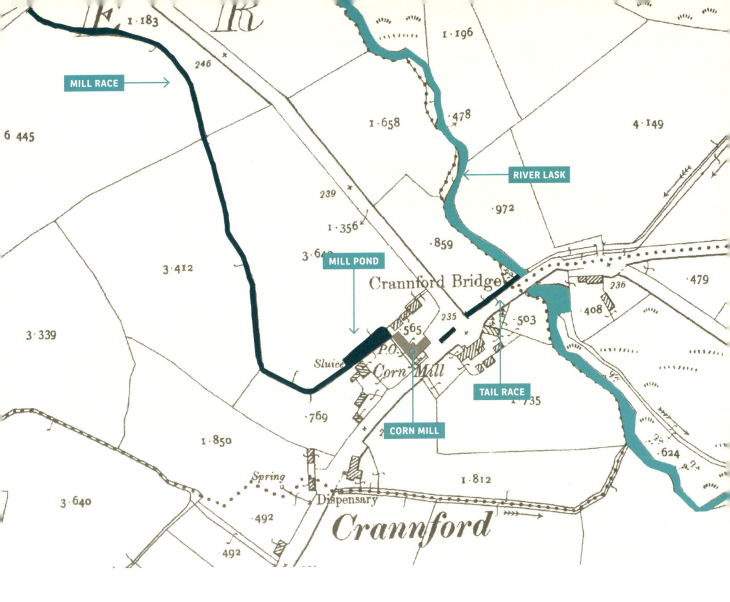

MILL RACE

6 445

1·183

246

1·196

·478

1·658

RIVER LASK

4·149

239

·972

1·356

·859

3·412

3·6

MILL POND

Crannford Bridge

·479

236

3·339

565

235

·503

408

Sluice

Corn Mill

TAIL RACE

·735

·769

CORN MILL

1·850

·624

3·640

Spring

Dispensary

1·812

·492

Crannford

·492

thanks to a God". The kiln room, where the kiln fire burned twenty four hours a day, was the hub of a community. The local people would meet there in the evening to discuss local matters and to find out from the travelling journeymen what was happening in the rest of the world.

A journeyman was a tradesman who travelled the country doing a few days work at any one place. He could be a thatcher, stonemason, carpenter, basket weaver or have some other trade which was needed in the community. When he came to a village he went directly to the mill because the miller knew where work was available in the farms in the area. The journeyman would come back each evening to the mill where he would sit with the villagers around the kiln fire, singing songs and telling stories and the news that he had picked up on his travels. He generally slept in the kiln room which was always warm. This was one way news got around in rural Ireland until the coming of more modern media.

CLOTH MILLS. Before the 1760s, textile production was a cottage industry using mainly flax and wool. The techniques of textile production had been known for centuries, and the manual methods had been adequate to meet the demands of the clothing industry. When Europe began to import cotton the balance of demand and supply changed.

FIG. 59
Bedstone for crushing oats,
Crannford Corn Mill (Don McEntee)

FIG. 60
Millstone for grinding wheat,
Crannford Corn Mill (Don McEntee)

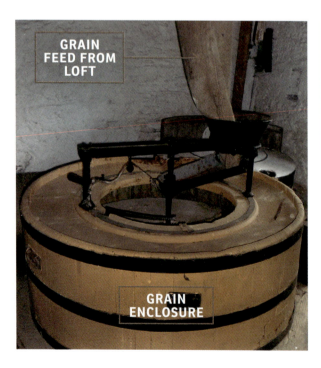

FIG. 61
Millstone for grinding animal feed,
Crannford Corn Mill (Don McEntee)

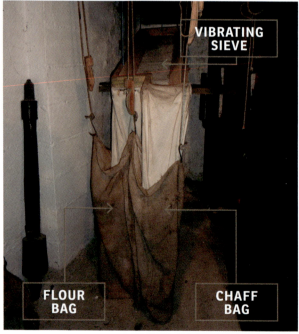

FIG. 62
Flour sieve separating flour from chaff,
Craanford Corn Mill (Don McEntee)

'Wash mill' cross section

1 'Foot' or 'stock'
2 'Shank'
3 Pivot
4 Massive wooden framing
5 'Heel clout'
6 Facings
7 Revolving wiper on the cam-shaft
8 Cam-shaft, driven from the waterwheel
9 The 'box' into which swing the 'feet' or 'stocks'
10 Drainage holes
11 Water cock
12 Retaining hooks, to hold back the 'foot' when not in use

FIG. 63 (TOP LEFT)

Water driven fulling stocks (Vittorio Zonca 1607)

FIG. 64 (TOP RIGHT)

Wash mill of Irish linen industry (W.A. McCutcheon)

Cotton spinning was patented in England in 1769. The preparatory processes and spinning took place in a spinning mill, weaving in a weaving shed and finishing at the bleach works and dye works. Traditionally these processes took place in separate mills. The English cotton mill, which emerged as an entity in 1771, went through many changes over the next two centuries and had a worldwide influence on the design of mills. The end of the patent in 1783 was followed rapidly by the erection of many cotton mills. Similar technology was subsequently applied to spinning worsted yarn for various textiles and flax for linen.

At the end of the eighteenth century hand finishing of cloth was replaced by tuck mills where water-driven beetling engines, wash mills and rubbing boards were the three main items of machinery required in the new bleach greens being set up for the cloth industry at that time.

From the medieval period, the fulling of cloth was undertaken in a wash mill, known as a fulling mill, or a tuck mill. Zonca's drawing, **FIG. 63**, illustrates the principle of water-driven fulling stocks and the details are shown in **FIG. 64** as used in the Irish linen industry. These mills were designed to cleanse, thicken and strengthen newly woven cloth. The cloth was beaten with wooden hammers, known as fulling stocks or fulling hammers. Fulling stocks were of two kinds: falling stocks (operating vertically) that were used only for scouring, and driving or hanging stocks. In both cases, the machinery was operated by cams on the shaft of a water wheel, or on a tappet wheel which lifted the hammer.

Driving stocks were pivoted so that the foot (the head of the hammer) struck the cloth almost horizontally. The stock had a tub holding the liquor and cloth. This was somewhat rounded on the side away from the hammer, so that the cloth turned gradually, ensuring that all parts of it were milled evenly. However, the cloth was taken out about every two hours to undo plaits and wrinkles. The "foot" was approximately triangular in shape, with notches to assist the turning of the cloth.

There are indications that water driven silk spinning mills operated in the Italian cities of Bologna, Florence and Lucca, in the fourteenth and fifteenth centuries. Zonca illustrated silk mills in 1607 **[FIG. 65]** and Leonardo da Vinci made drawings of a complete spinning machine around 1500.

PAPER MILLS. Paper was invented by the Chinese by 105 AD during the Han dynasty and spread slowly to the west via Samarkand and Baghdad. Muslims living on the Iberian Peninsula (present day Spain and Portugal) began papermaking in the tenth century. The process gradually extended to Italy and southern France, eventually reaching Germany by 1400. In medieval Europe, the hitherto handcraft of papermaking was mechanised by the use of waterpower, the first water driven paper mill in the Iberian peninsula being built in the Portuguese city of Leira in 1411. The rapid expansion of European paper production was enhanced by the invention of the printing press and beginning of the Printing Revolution in the fifteenth century.

In the paper-making process, raw material consisting of linen, cotton, rags and other fibrous material is reduced to a pulp in a beating or stamping process. Water-powered stamping mills, introduced in Spain during the fifteenth century, were used in making paper. A stamping mill consisted of three sets of stampers with four pestles in each set. Each pestle was lifted and let fall by the rotation of a camshaft. **FIG. 66** illustrates a stamping mill driven by an overshot waterwheel through a massive horizontal camshaft. The pulp, after being transferred to a vat, was tipped out of a mould, pressed and dried. When the sodden pulp was spread out over a wire mesh the water drained off, leaving the fibres to form the paper.

The paper was next dried and smoothed to give a satisfactory sheet, known as laid paper. On all handmade paper the impressions which can be seen in the paper are caused by the wires in the screen. A maker's paper can be identified by the firm's watermark which is created by weaving a design into the wires in the mould. In the 1780s the manufacture of woven paper spread to the paper mills in England and Ireland. The wire mesh in the mould was replaced with a woven brass wire-cloth cover producing a sheet unblemished by the furrows formerly found in laid paper; this is wove paper. Following the establishment of the paper machine in 1807, the manufacture of paper on a wove wire base never looked back. Today more than 99% of the world's paper is made in this way.

With the introduction of steam power in the nineteenth century the technology of mass paper production from wood pulp was developed. Paper

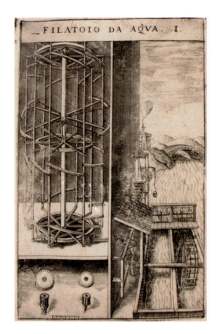

FIG. 65
Silk spinning wheel
(Zonca 1607)

FIG. 66

Water driven stamping
mechanism used to reduce rag
materials into usable fibres
for papermaking
(Diderot 1767)

made from cloth was far superior to that made from wood pulp. Dublin was a centre of excellence for the production of high quality paper in the eighteenth and nineteenth centuries in Ireland. The majority of the water-driven paper mills were located in the Dodder catchment.

James W. Phillips (1916-1986) wrote an account of the Dublin book trade in 1952 as a University of Dublin doctoral thesis. This was published in a book titled "Printing and Bookselling in Dublin 1670-1800" by the Irish Academic Press in 1998. Extracts dealing with papermaking in the Dodder catchment are included in this book.

The earliest official documents concerning the paper industry in Ireland were the petition and letters patent granted to Nicholas Dupin and the Company of White Papermakers in 1690 by the parliament of the Kingdom of Ireland. Dupin was probably one of the Huguenot paper factors who fled France after the revocation of the Edict of Nantes in 1685. He was granted a monopoly for fourteen years for the manufacture of white writing and printing paper. He also got a patent for the manufacture of linen and papermaking in England and Scotland.

Linen rags were used as raw material to produce good quality white paper. The first mill to produce paper was probably the one built on Lord Wharton's property in Rathfarnham. Mill No. 19 on the Owendoher catchment was making paper in 1693 with the Dupin watermark. Dupin set up the enterprise by organising the construction of the mill and providing foreign tradesmen to operate it. He did not have time to get involved in the day to day running of the mill due to his many other enterprises.

The second papermaking venture was mentioned in the will of Samuel Lee, a Dublin printer who died in 1694. Lee held a third share in the mill, which was situated at Milltown; the other partners were Captain Richard Price and Richard Wild. It is possible that this mill was constructed by Dupin's company and leased to Lee, Price and Wild. As Dupin held the patent for fourteen years, he would not have allowed a competitor to open a paper mill at that time. It is not known for how long this mill operated after Lee's death.

The next papermaking enterprise of which documentary proof exists was that of John and Edward Waters. A memorial of an Indenture of Lease dated 2nd February 1711 between Joseph Leeson of the City of Dublin and John Waters of Milltown, papermaker and printer and his son Edward Waters, whereby Joseph Leeson "did let to John and Edward Waters a House and Mill formerly a Tuck Mill but now a Paper Mill together with the Land Orchard Cabbins Water Courses Comiditys and Appurtenances whatsoever....lying and being near Milltowne Bridge in the County of Dublin."

To foster the papermaking industry in Ireland the Irish Parliament in 1759 imposed an import duty of "one shilling per rheam on all paper exceeding five shillings in value." This had the dual purpose of supporting and protecting the infant manufacturing industry of paper in Ireland and discouraging the importation of French and other foreign papers. In 1795 the duty was increased from one shilling per rheam to two pence per pound. The printers were not pleased with this increase because of the hardship it caused to the printing industry. Three years later they got Parliament to reduce the duty to one penny per pound. The efforts of the Irish Parliament to foster and stimulate the papermaking industry were not successful and after the Act of Union in 1800 there was no further help to the industry.

From its inception, the Royal Dublin Society fostered the arts and industries of Ireland. In 1747 it established a fund for annual awards for the products of the papermakers. Parliamentary grants for the construction of mills and collection of materials were made in the 1750s and 1760s through the Society's offices. Practically all the Irish papermakers received grants from one hundred to five hundred pounds from the Parliament. They also introduced a grant for the best raw material in 1759. This grant was the first evidence of the growing concern in Ireland about the matter of raw materials for the industry. Raw materials were the concern not only of Irish papermakers but of all papermakers throughout the Western Hemisphere during the remainder of the eighteenth century.

From 1770 through to 1800 there was a sharp decline in the texture and colour of all printing paper. During those three decades books printed in Dublin were frequently printed on blue coarse paper. This decay in quality is traceable to the raw materials shortage, for without the best liner rags the best paper cannot be made.

IRON MILLS. Prior to the Industrial Revolution every town or village had at least one forge. Here a blacksmith created objects from wrought iron by forging the metal; that is, by using tools to hammer, bend and cut. Blacksmiths produced objects such as gates, grilles, railings, light fixtures, furniture, sculpture, tools, agricultural implements, decorative and religious items, cooking utensils and weapons. His trade was an essential part of community life.

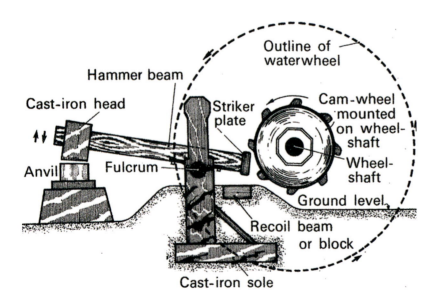

FIG. 67
German tilt hammer
(W.A. McCutcheon)

A blacksmith could be assisted by a striker whose task was to swing a large sledge hammer in heavy forging operations. The blacksmith directed his assistant by holding the hot iron at the anvil (with tongs) in one hand, and indicted where the iron was to be struck by tapping it with a small hammer held in his other hand. The striker then delivered a heavy blow where indicated with the sledge hammer. This forging shaped and strengthened the wrought iron into the shape of the required product. For smaller items the blacksmith himself wielded a lighter hammer.

Tilt hammers, introduced into the European iron industry from the seventeenth century onwards, were used in the mass production of wrought and cast iron parts. Tilt hammers were lifted by a cam on a shaft driven from a water wheel [FIG. 67]. FIG. 68 shows the interior of an iron forge (Diderot, 1765) with a massive tilt trip hammer being used to forge the bloom in the making of wrought iron. The shape of the hammer head determined the shape of the finished product.

From around 1780 water driven spade mills with tilt hammers [FIG. 69] began to appear in all districts of Ireland as local blacksmiths could not keep up with growing demand. FIG. 70 shows a range of spades at one of these spade mills.

SAW MILLS. The earliest recorded sawmill was the Roman water-powered stone structure at Hierapolis in Asia Minor (modern-day Turkey). Dating from the third century AD it is also the earliest machine known to incorporate a crank and connecting rod. By the eleventh century, water powered timber sawmills were in widespread use in the medieval Islamic world, from Spain and North Africa in the west to Central Asia in the east. They were also extensively used throughout medieval Europe.

Prior to the invention of the sawmill, boards were rived and planed, or more often sawn by two men with a whipsaw, using saddle blocks to hold the log, and a saw pit for the pitman who worked below. Sawing was slow, and required strong, hearty men.

Early sawmills simply adapted the whipsaw to mechanical power, generally driven by a water wheel to speed up the process [FIG. 71]. The circular motion of the wheel was changed to reciprocating (back-and-forth) motion of the saw blade by a connecting rod known as a pitman arm. Generally, only the saw was powered, and the logs had to be loaded and moved by hand. An early improvement was the development of a mobile carriage, also water powered, to move the log steadily through the saw blade.

The next improvement was the use of circular saw blades and soon thereafter, the use of gang saws which added additional blades [FIG. 72] so that a log would be reduced to boards in one quick step. Circular saw blades were extremely expensive and highly prone to damage from overheating or dirty logs.

Water driven sawmills were common in all parts of Ireland. They were used to produce timber planks from both soft and hard woods. FIG. 73 shows the interior of a saw mill at Ferrybank, Co. Waterford; where one drive shaft provided power, by means of drive belts, for different activities in the mill.

FIG. 68
The interior of an iron forge
showing massive hammer
used in forging cast iron
(Diderot 1767)

FIG. 69
Spade mill tilt hammer
(W.A. McCutcheon)

FIG. 70
A range of spade and
shovel types, Carnanee
Mill, County Antrim.
(W.A. McCutcheon)

FIG. 71 (TOP LEFT)
Saw mill
(Agostino Ramelli, 1588)

FIG. 72 (TOP RIGHT)
Saw mill
(Salomom de Caus, 1615)

OTHER MILL TYPES. An edge mill is a device which utilises a single or pair of round stones on their edges that revolve around a central pivot in a shallow, round milling bed [FIG. 74]. The edge rollers crushed rather than ground and were applied in the medieval period to the production of oil, sugar, pigments, tannin and mortar.

Polishing mills: water driven polishing and burnishing mills were used to sharpen, polish and burnish various forms of arms and military equipment [FIG. 75].

Finally, water wheels were also used to drive various items of equipment used in mining. Water trip hammers [FIG. 76] were used by the Romans to crush ore into small pieces for further processing. In the middle ages miners used water wheels to pump water from mines, grind ore, run bellows at blast furnaces and operate hammers in the forge.

FIG. 73
Murphy's saw mills,
Ferrybank, Waterford, 1901

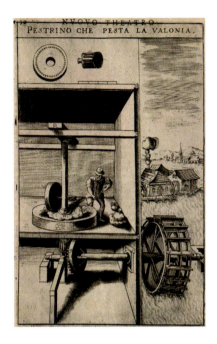

FIG. 74
Roller mill used in
production of oil, sugar
(Zonca, 1607)

FIG. 75
Waterwheel for sharpening
and burnishing arms
(Zonca, 1607)

FIG. 76
Water trip hammer
for crushing ore
(Zonca, 1607)

CHAPTER 9

ᎷMILL LISTS OF THE DODDER

DODDER MILLS AND MILLRACES

In 1991, in *The Rivers of Dublin* CL Sweeney observed: "The Dodder, though uncontrollable at times has been, on the whole, a useful servant." It was in fact one of Ireland's most productive industrial rivers. Together with its tributaries, it turned out an amazing range of goods, services and raw materials.

As to materials, the Dodder's usefulness in ancient times as a source of raw materials was noted in Rutty's *Natural History of the County Dublin* in 1772: "ragstones, for putting an edge to iron tolls; also clay, for making crocks and pantiles, abound on the shores of the Dodder from Old Bawn to Castle Kelly". In the seventeenth and nineteenth centuries, it was also a source of stone and gravel.

Just how intensively the industrial potential of the Dodder had been exploited by the mid-nineteenth century was made clear in 1844, when the Mallet report listed, numbered and described twenty-eight mills then existing on the Dodder, or relying on water from that river, beginning at Oldbawn. Wind, water and (incipiently) steam were the only energy sources available in the 1840s. Wind, a fickle and intermittent motive force, was not taken into account in the Dodder catchment. Steam, as yet only gaining acceptance in Ireland, was regarded as expensive because coal to heat the boilers had to be imported. Gas and electricity, which would eliminate the need for industry to be located beside a river, still lay in the future.

Most millraces and mill ponds were carefully planned and laid out. A weir was constructed in the main river channel diverting water into a millrace which led the impounded water along the side of high ground to a pond above the mill. An overflow from the mill pond would discharge excess water back to the river or to another mill race for the next mill. Generally, the mill pond was filled continuously from the river. When the mill started operations the water from the pond was fed down a sluice trough beside the building to a vertical waterwheel which turned a shaft inside the mill. Having performed its task, the flow of water was then fed back along a tail race into the natural course of the river a short distance downstream—or to another mill race. Maintaining a good flow throughout the year was essential.

Those who depended on rivers to power their mills or other industries enjoyed ancient rights which they guarded jealously, often long after other forms of motive power had superseded water. The Early Irish Laws recognised the right of a miller to acquire water from a river and construct a race to his mill across a neighbours land. Water rights frequently became the subject of bitter and expensive litigation and public bodies took care not to obstruct or otherwise interfere with supplies to millers

DODDER INDUSTRIAL LISTS

Mills along the Dodder started up and ceased over the centuries and it is difficult to reconcile the lists compiled at different times. In addition to Mallet's, a detailed record was compiled by William D. Handcock in 1879 and revised in 1899. Other lists and references can be found in the Dublin Historical Record (DHR), James Hegarty's exhaustive survey of the Dodder Valley and its landmarks (Vol. 2, 1939-40) being most useful. A list of the Dodder mills was also published in the DHR Vol. 12, 1951.

Archer's 1801 Survey lists three cotton-wash mills in Ballsbridge, owners Duffy, Byrne and Hamill.

Several other lists of milling enterprises powered by the Dodder have been compiled by various people at different times. Some of these record mills which were gone before Mallet's 1844 Report or which started up subsequently. Many of them are recorded here, but it is difficult to reconcile all the inventories and to account for all the mills. With additional information from *The Rivers of Dublin*, William Smith's 1879 catalogue, Domville Handcock's *History of Tallaght* and other sources quoted in the text, a list of mills on the Dodder follows, using Mallet's numbering as the principal record. Where possible, the other available lists are related to Mallet's work.

Mills changed hands and uses from time to time—some of them more than once—and so more than one proprietor's name can be found for a particular establishment. It should also be noted that some mills and other industrial enterprises were short-lived and that the use of several factories changed, sometimes more than once. And the information available about some of them is regrettably minimal.

Most industries cause pollution, which was not regarded as a problem when the first factories were established on or adjoining rivers in times long past. A noteworthy feature of the Dodder's industrial era, however, was the number of paper mills it sustained and which were notorious for the high levels of pollution they produce. This pollution was discharged into the local river and had a deleterious effect on fish life.

Many of the Dodder mills are of considerable historical significance and, in these instances, additional material from other sources has been added to Mallet's paragraphs. Mills and factories are listed by area, type and owner. Hopefully, this will enhance understanding of Mallet's and the other lists.

THE HEALY LIST

Because he was concerned specifically with the main courses of the Dodder and Poddle rivers, Robert Mallet's report omitted a number of local mills. Although these were close to the Dodder, they were driven by separate watercourses. In *The Mills and Mill-races of the Tallaght Area,* Patrick Healy described these mills in detail. Using material from his book, mills in the Tallaght area expand the Dodder list compiled by Mallet. To avoid confusion, each mill from Healy's list is identified by a letter rather than a number.

WILLIAM SMITH'S 1879 LIST

On 21st October 1879 William Smith of 41 Bloomfield Avenue wrote a letter to which he attached "a list of the Mills and Manufactories on the Dodder and the City Watercourse as also on Lord Meath's Watercourse." The letter and list were connected with some now unidentifiable legal process, with 45 numbered entries. The compiler was clearly uncertain as to the ownership or working status of some of the industries.

Smith's list included breweries, paper mills, foundries, tanneries and no fewer than 15 flour mills. His first five entries were situated upstream of the Balrothery Weir, while Nos. 6 to 13 were on the Dodder downstream of the weir. Nos. 14 to 45 were on the Poddle or one of its branches.

Allowing that 35 years separated Smith's list from Mallet's and that many industries came and went in the interval, it is still possible to reconcile some of the entries on one list with those on the others. Where the information coincides, this is duly noted in the appropriate paragraph. All entries are quoted in numerical order going downstream.

MILLS LISTED IN THIS BOOK

Most of the mills detailed in this book, other than those referred to in Mallet's Report, were generally those located on the Six Inch 1837 Ordnance Survey Maps of Dublin.

MAP NO. 26
Old Mill, Friarstown
(Duncan's map, 1821)

MAP NO. 27
Mill B – Parchment Mill,
Kiltipper (OS Map, 1912)

CHAPTER 10

ꟽILLS IN THE UPPER DODDER CATCHMENT

he River Dodder above the Balrothery (City) Weir was known as the Upper Dodder by The Rathmines and Rathgar Commissioners, a description given legal recognition in their 1880 Water Act.

MILLS AT KILTIPPER AND TALLAGHT

Mills B, C and D are noted on Healy's list. The water to power these three mills came from local streams which flowed into the Dodder.

MILL A OLD MILL FRIARSTOWN. The first mill on the Dodder is shown on Duncan's Map 1821 [MAP NO. 26]. It was probably a corn mill located on the right bank of the river upstream of the ford and footbridge where the Fort Bridge is now located. The mill race was probably fed from the nearby Friarstown Stream.

MILL B KILTIPPER PARCHMENT MILL. The millrace which served the Oldbawn Paper Mills (No. 1 on Mallet's list) was located on the Dodder at Kiltipper. Healy records this millrace as existing before the middle of the eighteenth century and was probably much earlier, as there is a large watermill shown at Oldbawn on the Down survey map of 1655. The uppermost mill on this millrace was a parchment mill at Kiltipper [MAP NO. 27]. In the Cobb Estate rentals of circa 1816 it was described as a good slated skin mill held by George Johnson, an additional note saying that it had been burned down.

The parchment mill was in ruins by 1837, although some slight remains still existed on a high bank above the Dodder, with the route of the mill race visible. Observing that it was difficult to see how the mill race could be fed by

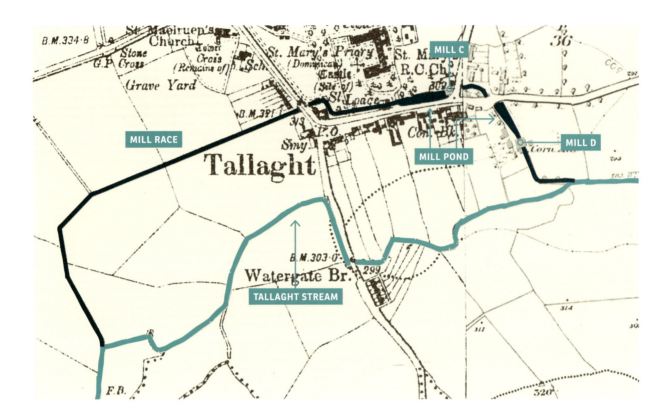

MAP NO. 28
Mills C & D Tallaght
(OS Maps, 1912 & 1843)

the Dodder, Healy noted that some slight remains of the mill survive: a high bank over the Dodder with the track of the millrace adjoining. Even allowing that the river bed has been deepened considerably in the last two centuries it is difficult to see how water from the Dodder was raised to this height, making Handcock's theory that the water came from the Ballymaice stream more likely.

MILL C DOMINICANS' WATER MILL The Dominican Priory occupies the site of the former Palace of Tallaght. In the 1835 OS Map [MAP NO. 28] there is shown a mill race flowing into a mill pond in the grounds of the Old Castle. The outlet from the pond flowed into the mill pond of Mill D.

In 1880 a small water mill was erected in the grounds to pump a domestic water supply to a tank on top of the tower. This installation was designed and built by Brother John Perkins who had previously worked as an engineer in Oldbawn Paper Mills. It continued to operate until 1950.

MILL D THE MANOR MILL. Known as the Archbishop's Corn Mill [MAP NO. 28], this was located on present day Greenhills Road between Tallaght Bypass and Main Street. The tenants of the Archbishop's Manor had their corn ground at this mill, described in 1837 by O'Curry as the smallest and oldest he ever saw with two pairs of stones capable of grinding only about four barrels of wheat per day. Eugene O'Curry, with John O'Donovan, was one of the greatest nineteenth century scholars of antiquity.

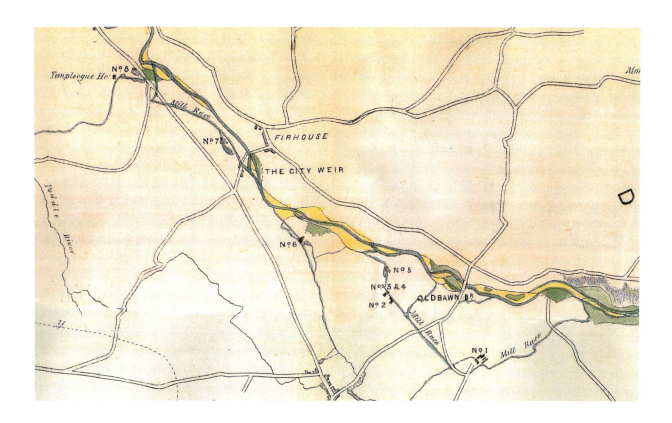

MAP NO. 29

Upper Dodder Mills 1 to 8
(Mallet's Map, 1844)

The mill was operated by an overshot wheel five feet in diameter and two feet nine inches wide, and the fall of water was eleven feet. According to Archer's survey of 1801 the miller was Newman and subsequent directories gave the name of Michael Mahon down to 1849. The walls which carried the mill wheel survived until 1980 when the watercourse was culverted. This mill ground the corn of all the manor's tenants under the usual toll or mulcture of one shilling for every barrel not processed there.

In the grounds of the Priory there is a millstone which probably came from this mill [FIG. 77].

Healy notes that, after servicing the Manor Mill, the millstream was conveyed across the fields (present day Tallaght Stream) to Bolbrook Paper Mill (No. 6 on Mallet's list). Here, on account of its purity, it was used in the manufacture of paper.

FIG. 77

Millstone in garden of Tallaght
Priory (Don McEntee)

MILLS ON UPPER DODDER

The mill numbers on the Dodder are those used in Mallet's Report except mills numbered 27[1], 27[2] and 28 which have been renumbered 27, 28 and 29. Mills on the main tributaries are numbered separately. Location of Mills Nos.1 to 8 is shown on MAP NO. 29.

MILL NO. 1 OLDBAWN PAPER MILL . [MAP NO. 30]. Domville Handcock noted that "at little below Friarstown" (at a point called Diana or Kiltipper), "nearly the whole of the water is here diverted into the mill-race by an ill-constructed weir of loose stones and sods requiring renewal after every flood. This diverts most of the river into a millrace, which for some distance follows its course under the left bank". This same millrace also served the next five mills. Handcock recorded that the millrace was carried in a very crude channel, full of leaks and overflows and was staunched, as occasion required, with sods and boards. When a flood came down, the weir was swept away and all the mills below as far as Firhouse were left idle until repairs were carried out.

In the 1740s Michael McDonnell built a new paper mill at Tallaght. In his petition to the Irish House of Commons in 1757 he gave an outline of this experience in the manufacture of paper. He stated therein that he had spent long years abroad studying the industry and had worked in mills in the Netherlands and France. With Michael McDonnell there began a dynastic family of papermakers that was to continue into the twentieth century, a longer period than any other in the Irish industry. In addition to Michael, there were John, Matthew, Christopher and Richard McDonnell active in the paper trade in the Dublin area during the eighteenth century.

Healy further records that Oldbawn House had been bought shortly before 1800 by Joseph McDonnell who built the extensive paper mill on an adjoining site. The north wing of the old house was completely cleared out of its internal features from ground floor to attic and incorporated into the mill. The windows were broken into large vertical openings and fitted with latticework to admit air which dried out the paper. The McDonnell family occupied the rest of the house.

The Oldbawn mill, steam-powered by 1876 with one 200 hp engine and some smaller ones, supplied several of the Dublin newspapers but closed around 1878 due to foreign competition and was abandoned by 1899. It is shown on a Waterworks Map as at one time belonging to McDonnell. Healy states that the north wing of Oldbawn House, used for drying out paper, lay derelict until 1976 when it was demolished.

The *Dodder Valley News* recorded that a large watermill at this location was later converted into a paper mill by the McDonnell family. No. 1 on William Smith's 1879 list, the Old Bawn Paper Mill Company is quoted as the lessee of the mill, stated to be not working; the owner was Joseph Pratt Tyute, represented by Mr. Dunne, solicitor.

MAP NO. 30 (OPPOSITE)
River Dodder Mill 1
(OS Map, 1912)

MAP NO. 31 (RIGHT)
River Dodder Mills 2 to 6
(OS Map, 1843)

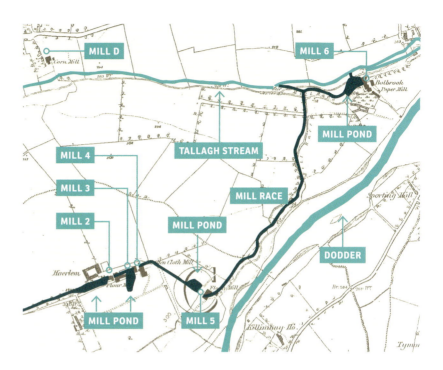

MILL NO. 2 MCCRACKEN'S FLOUR MILL.

MILL NO. 3 FLOUR MILL.

MILL NO. 4 WOOLLEN MILL. Mills Nos. 2 to 6 are shown in MAP NO. 31. Healy states that there were four mills at or near Oldbawn around 1836, the first being worked by McCrackens until 1863, then by Thomas and John Neill. This mill was burned down in 1887. There are slight discrepancies between the accounts of the various historians, but this is possibly due to the use, ownership and occupancy of premises changing. Handcock refers to four mills in this area, all owned by Messrs. Neill—one woollen and three flour mills. Nos. 2, 3 and 4 were flour mills, all owned by Joseph Fade Hutchenson, occupied by Thomas and John Neill.

MILL NO. 5 NEILL'S WOOLLEN MILL. In 1776, this mill had been worked by Haarlem and Company, calico printers. Here, Handcock wrote, was one of the most celebrated bleach greens in Ireland. He also recorded that, in 1813, one of the principals of this business was a member of the philanthropic Bewley family, whose benevolence alleviated local poverty.

Below this was a flour mill and adjoining woollen mill owned by Manus Neill which passed to Thomas and John Neill in 1871. The premises ceased to be used around 1886.

Healy points out that the mill described as Neill's Woollen Mills by Mallet in 1844 was reported a few years later to be a flour mill. It was held by Thomas and John Neill in 1879 but had closed down before 1900.

MILL NO. 6 WILLIAMSON'S (BOARDMAN'S) BOLBROOK CARDBOARD AND PACKING PAPER MILLS. Also referred to as a pasteboard. Healy records that in 1760 Pierce Archibald, a carpenter, built a couple of two-vat mills here and employed 30 people. Following a fire in 1762 which destroyed the entire complex and cost £2,000, the mill was replaced but on a lower site on the opposite side of the road, probably to take advantage of additional water power; this new mill had a staff of 50. It was sometimes referred to as Newtown Mill. Pierce made paper at these mills until his death in 1776. He was succeeded by John Archibald, a son or relative. Upon John's death in 1788, the mill passed into the hands of Elizabeth Archibald, the widow of John. In Archers Survey of 1801 the owner is quoted as Widow Archibald, the premises being later let to J. Williamson.

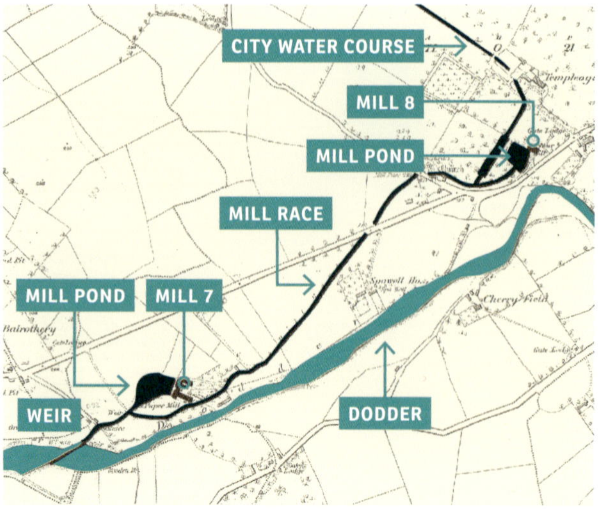

Around 1854 this mill was leased by Anne Archibald to Batter & Co., later changed to Batter and Boardman and subsequently Thomas Boardman and J. Williamson before passing to the Boardmans—first Thomas, then Adam (1875) and Robert (1893). Adam was the owner in 1875, Robert in 1893; Robert occupied the mill until 1904 but by 1908 it was vacant and dilapidated. The buildings, used to house farm animals by 1910, were later cleared in the landscaping of the Dodder banks.

After servicing Bolbrook Paper Mills, the tail race returned to the main channel of the Dodder, joined also at this point by the Tallaght Stream. The site of the mill was on the open ground between the Tallaght bypass and the Dodder, now part of the Bolbrook Estate.

MILL NO. 7 BELLA VISTA [MAP NO. 32]. The millrace to this mill started at Balrothery Weir (City Weir) and continued on to Mill No. 8 and the City Watercourse.

James McDonnell's paper mill had the largest water wheel of all the Dodder mills. It was 24 feet (7.3m) in diameter and eleven feet (3.35m) wide and worked at a speed of four revolutions per minute.

In the 1719 Act of Parliament relating to the City Watercourse, this mill is referred to as Ashworth's new Delaford paper mills. The new mill was erected by Daniel Ashworth 'of the city of Dublin Currier'. That same year he mortgaged the mill to William and Jacob Walton. In 1720 he made over the lease to William Barry. Ashworth, however, continued to operate the factory. In 1721 he presented a petition to the Irish House of Commons for assistance in the enterprise. Ashworth declared that there 'has been great expense in building and erecting a mill, and procuring other materials for the making of Paper fit for either Writing or Printing, and have brought the same to a handsome perfection'. His petition was unsuccessful.

In 1722 an advertisement appeared in *Dublin Intelligence* for the sale of Ashworth's mill: "A very good large new Double Paper Mill, with 2 Vatts and a large Dwelling House, with a store House and several out houses all in extraordinary good Order ....to be sold." This indicated that Ashworth

MAP NO. 32
River Dodder Mills 7 & 8
(OS Map, 1843)

had died. His widow still had an interest in the property in 1730 when she and Barry's widow leased the Ashworth mill to Thomas Slator, papermaker, as an independent manufacturer. It is possible that he previously operated the mill for the Ashworth/Barry interest.

Thomas Slater applied for Government aid in 1733, appealing again in 1737 when he claimed to be the only white papermaker in Ireland supplying white paper for the book printers and for writing. He had sent artisans abroad to study the methods and processes used by continental papermakers. He said that he had two mills at Templeogue and had another planned. He was granted five hundred pounds to build a mill in 'the best Dutch Manner'. In 1751 the mill was noted as producing more paper than any other mill in Ireland.

From 1737 Slator assumed the position of master of the paper industry in Ireland. He received many awards from the Royal Dublin Society for the best paper of varieties 'of Irish Make' during the late 1740s and 1750s. He continued to be the most eminent papermaker until his retirement in 1765.

Edward Burroughs, nephew of Thomas Slator, assumed control of the mill in December 1765. William McDonnell of Lucan operated the mill from 1803.

A new owner, Joseph McDonnell, installed a steam engine in 1836 but the mill ceased operation in 1876. McDonnell was related to other paper makers, notably at Saggart and Killeen. In *The New Neighbourhood of Dublin* (1949) Joseph Hone and Maurice Craig refer to a building in the grounds of Bella Vista, by then demolished. They state that this building was an engine house for a paper mill and in the revision of the 1949 book, updated by Maurice and Michael Fewer in 2002, this building is reported to have been renovated as a dwelling house.

In the *Dublin Saturday Magazine* for 1866-67 (Page 647), referring to Firhouse, is the line "a paper mill of Mr. McDonnell's is near the stream" The Owendoher, also known as the Rathfarnham Brook, joined the Dodder below this mill.

MILL NO. 8 FRANCIS BURKE'S FLOUR MILL (TEMPLEOGUE MILL E HEALY LIST) [MAP NO. 32]. A mill is shown at Templeogue on the Down Survey map of 1655 and milling on this site is mentioned as far back as 1394. It was held by the Burkes early in the nineteenth century but was burned down in the mid-1800s. Entirely rebuilt, it passed to JC Colville and around 1879 to McGonchy & Co. The tailrace of this mill supplied ponds in the grounds of Templeogue House. It subsequently closed down but extensive remains survived until 1985 when work on the Tallaght bypass necessitated their demolition.

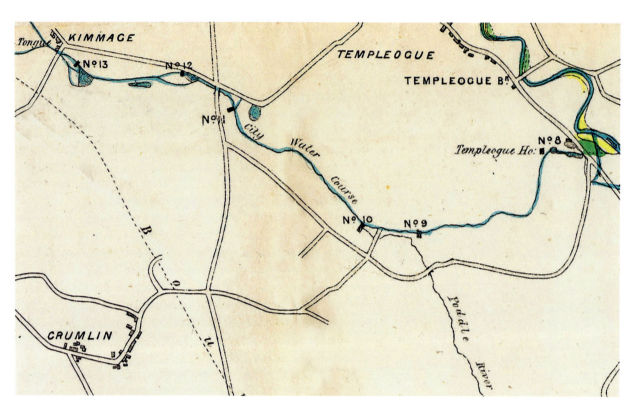

CHAPTER 11

MILLS AND INDUSTRY ON THE PODDLE

allet's mills numbered from 9 to 20 were located on the City Watercourse and the Poddle River. The City Watercourse started as the tailrace to Mill No. 8. As detailed on **MAP NO. 33** it joined the River Poddle in the vicinity of Mill No. 10 in the townland of Kimmage. The Poddle continued as a single water course as far as the Tongue (just downstream from today's Sundrive Road), where it divided into two water channels as shown on **MAP NO. 34**. With the increased and regulated flow in the lower Poddle that followed the construction of the Tongue, a milling industry developed in the vicinity of the city.

This necessitated a further diversion of the Poddle, another artificial watercourse being constructed from near Mount Jerome. Following a natural slope past today's Griffith College, this was built as a loop that rejoined the old stream at New Row. This diversion became known as the Abbey Stream. As the area within the Liberty developed, the higher and lower levels of the watercourse were linked by three smaller channels, called the Tenter Water, the Factory Water and the Commons Water — and subsequently altered to meet changing requirements. **MAP NO. 19** shows the interlinked channels within the city.

The City Weir at Balrothery that supplied the Poddle and City Watercourse was described in Mallet's report as being in a ruinous condition. Depending on an adequate supply of water from the Dodder for their operation, the establishments numbered 9 to 20 by Mallet were along the Poddle. The local information and comments in the report provide a glimpse into part of industrial Dublin in the mid-1800s.

Healy records that there were mills on the City Watercourse in the fourteenth century and refers to some establishments which did not appear on Mallet's 1844 list.

MAP NO. 33
River Poddle Mills 9 to 13
(Mallet's Map, 1844)

MAP NO. 34
River Poddle Mills 13 to 20
(Mallet's Map, 1844)

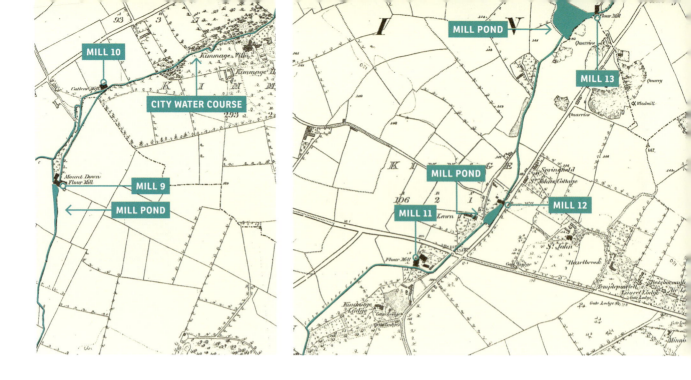

MAP NO. 35 (TOP LEFT)
River Poddle Mills 9 & 10
(OS Map, 1843)

MAP NO. 36 (TOP RIGHT)
River Poddle Mills 11 to 13
(OS Map, 1843)

In William Smith's 1879 list, entries Nos. 6-45 were on the Poddle or City Watercourse. He mentioned difficulties encountered after entry No. 31 "as the pipes and watercourses separate and join again so often and in such manner that I was unable to follow them". Here he was clearly frustrated by the number of branches and diversions that proliferated on the Poddle system over the years.

MILL NO. 9 MOUNT DOWN MILL [MAP NO. 35]. This was originally a woollen mill. The 1719 Act which sought to protect water quality outlawed tuck mills and woollen mills on the Poddle and City Watercourse, forcing the conversion of Mountdown to a flour and woollen mill. It was operated by the Sharpe family, but was taken over by the Cullens before 1900.

At the time of Mallet's report, the water reached this mill on an embankment "….in bad order….breaches frequently occurring….In flood times the fields all round this mill, to the extent of, perhaps, twenty acres or more, are liable to be flooded."

One of a series of flour mills, No. 15 on Smith's list is given as being leased by Sharpe (deceased) and it would appear to correspond with this entry by Mallet.

Under a High Court judgement in 1952 Mr. P. Cullen received £20,000 in compensation from Dublin Corporation for the loss of a water supply to the mill on the grounds that they had interfered with the water supply by lowering the crest of the City Weir. Cullen had used the water to generate electricity and drive a saw mill and corn crusher. When this section of the watercourse was filled in 1973 the mills and mill house were demolished.

MILL NO. 10 MILLIKEN'S CUTLER'S (BLADE) MILL. [MAP NO. 35] At the overfall above this mill a 3" (75mm) diameter pipe diverted water to supply Terenure House and Demesne. This is one of many nineteenth century instances of a domestic water supply being taken from a seriously contaminated source.

MILL NO. 11 TROY'S FLOUR MILL. [MAP NO. 36] As at Mountdown Mill (No. 9), the land around the embankment leading to this mill was also liable to flooding.

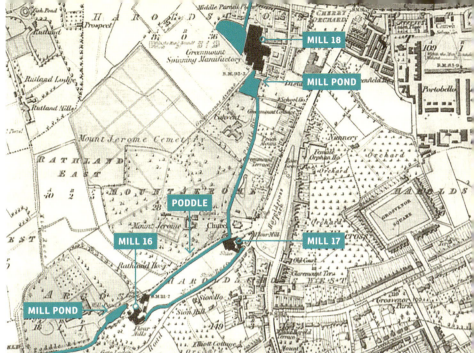

MAP NO. 37 (TOP LEFT)
River Poddle Mills 14 & 15
(OS Map, 1897)

MAP NO. 38 (TOP RIGHT)
River Poddle Mills 16, 17 & 18
(OS Map, 1897)

MILL NO. 12 REGAN'S FLOUR MILL; [MAP NO. 36]

MILL NO. 13 TUITE'S FLOUR MILL. [MAP NO. 36] This mill was located a short distance upstream of the Tongue, near Dark Lane (also called Hangman's Lane), the present Sundrive Road. The level of the mill pond and race serving Tuite's mill caused occasional flooding.

Pursuing the City Watercourse to the west from the Tongue [MAPS NOS. 19 & 34], the next mills were:

MILL NO. 14 GUY'S IRON MILL, makers of coach axles and boxes. The lessee of entry No.20 on Smith's list is given as Robert Guy, the premises being owned by a Mrs. Bond. On MAP NO. 37 in 1897 it called Rutlands Mills. It ceased to be a mill in early twentieth century.

MILL NO. 15. dilapidated and unoccupied, formerly tenanted by Guy (No. 14) and located at Dark Lane (now Sundrive Road). It is listed as an iron foundry by Smith in his entry No. 21, who gives the owner as Mrs Bond. This and the preceding iron works are given as on the City Watercourse "previous to its reaching the canal." On MAP NO. 37 it is shown as Ruland Flour Mill. In early twentieth century it became a laundry.

The Mallet report continued: "Going below this, the stream passes partly under the Grand Canal, in a culvert, near Camac Bridge, thence along Dolphin's Barn, and winding in an open channel at the rere of the Marrowbone-lane distillery, again passes under the Grand Canal by a second culvert at this spot and discharges direct into the city basin, at James's-street".

"Some time since, it appears the first, or upper, culvert became stopped with silt, as there is reason to believe it is to a great extent even now. An over-fall was then made from the higher or receiving end of the culvert into the Grand Canal; and since that, and now, a large proportion of the water brought down by the Dodder stream falls directly into the Portobello branch of the Grand

FIG. 78
Mill 16 at Mount Argus
(Frances McEntee, 1950)

Canal at this spot before ever reaching the basin directly. The remainder finds its way through the culvert, and passes by the old channel, which has lately been partially arched over into the James's-street city basin".

"The water introduced at present by the Dodder, for supply of the city, is turbid in floods, and from want of proper protection, defiled in various ways".

"If proper measures were adopted to convey it unpolluted into the basin, there would be, after the completion of the works proposed in the present report, such an increased delivery of water as would probably enable the basin to be supplied without any assistance from the canal, and thus supply the city from it with water of a purer quality than it receives at present."

Going downstream from the Tongue along the Earl of Meath's Watercourse, Archer's Survey in 1801 records eight mills in Harolds Cross :- one Corn Mill owned by Mr. Hyland, one Calender Mill owned by Mr. Armstrong, two Paper Mills and one Wire Mill owned by Mr. Cuppaidge, one Skin Mill owned by Wall Cogan Murphy (possibly rabbit skins), and two Corn mills owned by Bermingham Murphy and Co.

Joe Curtis in his book *"Harolds Cross"* gives a detailed account of the various mill in Harolds Cross.

Mallet listed the following:

MILL NO. 16 HICK'S PAPER MILL [MAP NO. 38]. FIG. 78 shows the mill circa 1950. This mill was known as Loader's Park Mill (formerly Lowther's Park). The first recorded owner of a mill here is Maurice Coffey. John Anderson, millright, leased the property for 55 years in 1719. Roques Map of 1760 shows a Corn Mill here.

After Anderson the next occupier was Hugh Kerr, followed by George Cuppaidge (paper and wire mills), Laurence McDonnell, John Cahill/Daniel Laffan (paper manufacturers), and Charles Lett. The mill continued as a paper

FIG. 79

Mill 17, Harold Cross Laundry
(Rathmines News and Dublin
Lantern, 1896)

mill until about 1882, after which it became a flour mill. Fanagan Funeral
Undertakers bought the mill in 1882 and used it as a stable for their horses.

The mill was capable of producing one ton of paper a day, suitable for
newspapers and printing. There was one large mill pond and one large settling
pond for the sludge arising from paper making.

The mill had two breast shot water wheels, one sixteen feet in diameter
and two feet ten inches wide and the other twenty feet in diameter and ten
inches wide. From the mid nineteenth century it also had steam power. The
combination of water and steam power meant that the mill could operate when
the flows were low in the Poddle.

MILL NO. 17 PETER MURPHY'S FLOUR MILL [MAP NO. 38]. This mill at the junction
of Mount Argus Road and Lower Kimmage Road, opposite Mount Jerome
Cemetery, operated as a flour mill for most of the nineteenth century. A Lord
Meath map of 1787 shows a mill and waterwheel. Joe Curtis quotes an 1850
Ordnance Survey House Book which described its water wheel, fourteen feet in
diameter, as having a fall of twelve feet and revolving six times a minute. The
breast shot water wheel powered two pairs of grinding stones, one four feet
eight inches diameter and the other four feet two inches diameter.

In 1894 the mill was converted into a laundry. It was known as the Harolds
Cross Laundry Company and Carpet Cleaning Works [FIG. 79] serving business
and domestic customers. The laundry continued in business until about 1970.
The site is now occupied by apartments.

MILL NO. 18 GREENMOUNT SPINNING COMPANY [MAP NO. 38]. In the 1760's there
was a corn mill on this site, called The Wood Mill leased from the Earl of
Meath. By 1799 the mill was owned by the Greenville Company. In 1807 James

Greenham bought the property and built a cotton mill, which later became known as the Greenmount Spinning Manufactory. Thomas, Joseph and Jonathon Pim, cotton merchants in South William Street, took over the mill in 1816. Important events included the installation of power looms during the 1830s, reducing the workforce (consisting mainly of teenage girls) from 300 to 150 and steam augmenting water power during the 1850s. Various extensions included a four-storey building in the 1860s.

The River Poddle drove a huge twenty two feet diameter by eleven feet wide waterwheel, which revolved 4.5 times a minute, had a fall of nineteen feet, while producing 25 horse power for nine months of the year , and 12 hp for three months of the year. When the Greenmount Weaving Company, as it then was, went into liquidation in 1922, 600 workers, mainly men, lost their jobs. The business was acquired by the Boyne Weaving Company of Drogheda in 1925 and formed the Greenmount and Boyne Linen Company [FIG. 80]. It finally closed in 1960, when the factory became the Greenmount Industrial Estate.

No. 28 on Smith's list—a cotton mill—corresponds with this factory. The owner is given as the Earl of Meath, with Johathan Pim as lessee.

The tail-race from Mill No. 18 discharged under the Grand Canal and passed the garden of the Richmond Penitentiary Prison (now Griffith College) through the Greenville dye-works. Here there formerly existed a fall and water wheel, which had been destroyed some time prior to Mallet's report.

Mallet's description of the Poddle and its many branches continued to operate until the 1960s, when it became the Greenmount Industrial Estate.

"A little below this (Greenville) at the site of the ancient Donore Castle the stream is divided—one part, passing off through the fields at the western side of Love-lane, across Cork-street, passes through the Marrowbone Lane distillery, after touching closely upon the city water-course again; and lastly, returning at the rere of Cork-street, and by various channels which, partly open and partly covered, runs along the streets of the Liberty, known as Pimlico and Tripoli, and is called, locally, the Pimlico river; while the branch which again meets this at Ardee-street, having passed at the rere of Cork-street, is known as the Poddle." This branch of the Poddle is also known as the Tenter Water.

MILL NO. 19 HAIGH'S IRON FOUNDRY: [MAP NO. 39] Mallet continued "At this spot the next fall exists (No. 19) and is occupied by Haigh for an iron foundry; he derives his main supply from the Poddle branch and has only an over-fall supply from the Pimlico stream, under which the former here passes at right angles in a culvert to the west side of Ardee-street."

The street layout in this area as it is now is very different from what was in existence (or not) in Mallet's day. This factory, designated as Samuel Fairbrother's wheelwright factory in 1864, was on a site that is now part of O'Curry Road.

"Returning again to the stream a little below Greenville works, and pursuing the right branch, we find it passes along Love-lane (now Donore Avenue) through Anderson's ancient calico print works; turns off to the east before reaching Brown-street, and passes on to the mill-pond of mill No. 20:

FIG. 80
Mill 18 Greenmount & Boyne Co., Harold Cross (Morgan Collection, 1950)

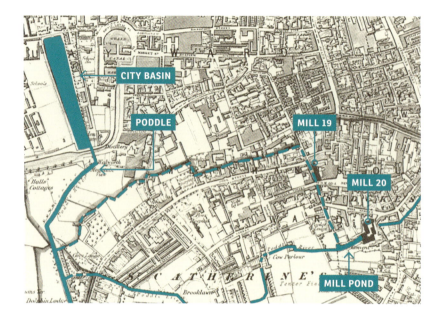

MAP NO. 39

River Poddle Mills 19 & 20
and River Poddle discharging
under Grand Canal in the city
basin (OS Map, 1897)

MILL NO. 20 BUSBY'S FLOUR AND OIL MILL. In Mill-street [MAP NO. 39]. "Into this mill-pond the waters of the Pimlico stream also discharge." This site was in later years occupied by the Warrenmount Mills (Ice and Cold Storage). One of the last short open sections of the Poddle in the inner city area lies along part of the former mill pond that served this premises.

Mallet continued his narrative with some general comments about the Poddle: "This is the last fall upon the river, whose waters, or at least, all that has arrived of those taken up at Templeogue, find their way from this point and from below Haigh's foundry (on the Poddle branch proper) in covered channels, known in the district by the name of the 'Poddle Hole' and passing near St. Patrick's Cathedral, through Little Ship-street, the Castle-yard, and across Dame-street, are finally discharged into the Liffey, under Wellington-quay".

"Along almost every part of this long, devious and intricate artificial course, or rather courses, much of which is above the level of the surrounding country, losses of water occur by partial seepages, leakage, breaches in drains, or small embankments, and wilful diversions of water, and it is subject to divers pollutions".

"...it would be proper that the whole of these channels should be cleansed and opened, and several minor repairs &c, executed; and a future surveillance exercised as to the unlawful abstraction of water."

Several factors led to the demise of industrial activity along the Poddle. Flour mills decreased in number following the abolition of the Corn Laws in 1846. While the modernisation of industry brought about by steam and, later, electricity, largely removed the necessity for water wheels—or for industries to be sited beside rivers. Many small factories, often worn out, which were producing obsolete goods or unable to compete with modernised production methods and cheap imports, closed down.

SMITH'S LIST — REMAINING ENTERPRISES ON PODDLE

The rest of the factories or mills listed by William Smith and which have no corresponding entries in Robert Mallet's list are catalogued below. Smith's numbers are used, prefixed by the letter S and where possible, are related to identifiable premises on the Ordnance Survey maps. Nos. S14 to S19 were on the City Watercourse upstream of the Tongue. Some of the streets on which these former mills stood have been substantially changed or redeveloped.

S14 Card board mill, owner Sir Charles Domville, occupier Samuel Wall.

S15 Flour mill, owner Sir Charles Domville, Lessee James C Colville, Occupier William McCouchy & Company.

S16 Flour mill, Owner Sir Charles Domville, Lessee Representatives of Sharpe (Decd), occupier William Dunford

S17 Flour mill, owner Sir Robert Shaw; occupier Anthony Greavan

S18 Flour mill, Owner Sir Robert Shaw; John Lennox

S19 Flour mill, Owner Sir Robert Shaw; Lessee and Occupier William Dnnford

S20 Flour mill, owner Mrs Bond; Lessees and occupiers Clibborn and Shaw—see Mallet No. 15

S21 See Mallet No. 21

S22 Tannery, Owner Miss Juliana Sharmon, Lessee and Occupier Charles Trench

S23 Tannery, Owner Miss Julianna Sharmon; Lessee and Occupier Laurence Byrne

S24 Tannery (not in working order). Owner Miss Julianna Sharmon; Lessee and Occupier Delia Jones

S25 Flour Mill (Loder Park). Owner, Captain Harvey; Lessee and Occupier James Whelan

S26 Flour Mill, Harold's Cross; Owner and Occupier LD Mulhall

S27 Cotton mill, Owner the Earl of Meath; Lessee Jonathan Pim, Occupier Greenmount Spinning Company

S28 See Mallet No. 18

S29 Glue Manufactory, Owner the Earl of Meath. Lessee Verschoyle, Occupier Verschoyle & Son

S30 Wood Turning Mill, Owner the Earl of Meath; Lessee and Occupier Samuel Fairbrother. The area in which this factory stood has been extensively developed and is now part of O'Curry Avenue.

S31 Dyeing Works, Owner the Earl of Meath; Lessee and Occupier Richard Eustace

S32 Distillery, Owner the Earl of Meath; Lessee and Occupier William Jameson & Company

S33 Belview Corn Mills, Owner the Earl of Meath; Lessee Alphonse Busby, Occupier Kate Ennis and Mary Ennis. This mill was located on the western side of Taylor's Lane on a site that is now part of the Guinness Brewery complex.

S34 Summer Street Brewery (formerly Caffrey's Brewery); Occupier Isherwood. This premises had frontages on to Summer Street and John Street South.

S35 Bacon Curing Establishment, Owner the Earl of Meath; Lessee and Occupier, Donnelly

S36 Brewery (Watkins); Owner Earl of Meath; Occupier Darby

S37 Malt House (New Market). Owner Earl of Meath Lessee and Occupier Alphonso Bushy

S38 Malt House (Mill Street); Owner Earl of Meath; Lessee and Occupier Reps of Read deceased

S39 Tannery (Mill Street), Owner Earl of Meath; Lessee and Occupier Patrick Cullen*

S40 Tannery, Mill Street; Owner Earl of Meath; Lessee John Molloy; Occupier John Molloy and E & C O'Callaghan*

S41 Tannery (Mill Street); Owner Earl of Meath; Lessee and Occupier Hayes Brothers*

S42 Malt House (Byrne's Hill); Owner Earl of Meath; Lessee and Occupier Peter Cahill

S43 Mill (Warren Mount Mill); Owner Earl of Meath; Lessee Alphonso Bushy, Occupier Henry Roe. Adjacent to Warrenmount Convent, this premises appears on the 1908 Ordnance Sheet XVIII.76 as in use for "Ice and Cold Storage"

S44 Brewery (City of Dublin); Owner Earl of Meath; Occupier City of Dublin Brewery Company

S45 Saw Mill (Coombe), Liberty Mill; Owner Earl of Meath; Lessee and occupier John Grace

Although mills declined in numbers over the years the rights of the remaining mill owners still had to be respected and a supply was maintained solely for their benefit. The Waterworks Department of Dublin Corporation inherited the responsibility of maintaining a supply from the headworks at Balrothery to the few remaining mills. On closure of these mills and forfeiture of rights the water supply from the Dodder was discontinued. The Poddle now drains the surface water from its natural catchment. Due to the extensive development in the catchment an overflow has been constructed through

* On the 1:1056 Ordnance Sheet No. 26 (1864) three tanneries are shown on the north side of Mill Street, one on the south side.

Terenure College to the River Dodder and a second overflow into the Grand Canal Sewer at Harold's Cross.

In 1861, the Corporation purchased the water rights of the Earl of Meath for £6,000, ensuring that nobody except the municipality could henceforth abstract water from the Poddle. However, the Earl retained the actual lands and the riverbed and banks where it flowed through his lands, and his right to use the water for operating mills also survived. The Liberty Basin or reservoir built around 1820 for the Earl at Pim Street and built over in Victorian times was uncovered in 2006 when the site was being cleared for redevelopment.

Throughout its long history as a water source, the Poddle saw several generations of mills come and go: at one time, there were nine such establishments in the section between New Street and the Liffey. Today, the Drainage Division of Dublin City Council maintains the Poddle network of surface water conduits.

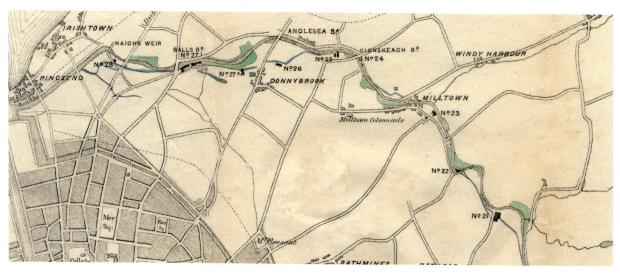

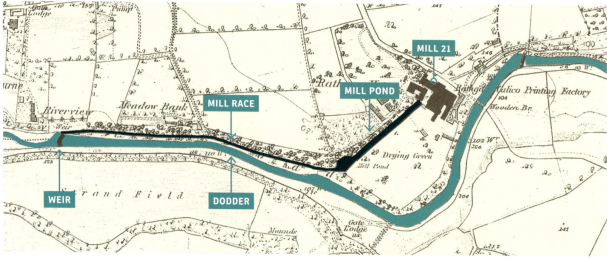

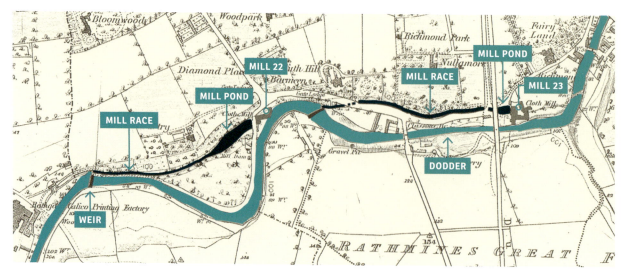

CHAPTER 12

MILLS ON LOWER DODDER (EAST OF BALROTHERY)

allet's list on the Dodder east of Balrothery includes mills Nos. 21 to 28 [MAP NO. 40]. The two mills numbered 27[1] and 27[2] were linked together as they were both owned by John Duffy & Sons. In this chapter the mills 27[1], 27[2], and 28 have been renumbered 27, 28 and 29. Sweeney recorded six major millraces with weirs and sluice water controls between Rathfarnham and Ringsend. He calculated the total length of these at 12 kilometres (3.6 miles), some of them servicing more than one enterprise. Two further millraces from contributing streams also discharged into this section of the Dodder.

MILL NO. 21 WALDRON'S CALICO PRINT WORKS (disused in 1844) stood downstream of the Rathfarnham Brook. Its millrace [MAP NO. 41 AND FIG. 81] started at a weir east of Rathfarnham (Pearse) Bridge, paralleling the north bank close to the river. After crossing into the (later) High School playing fields it flowed into a long millpond that served Waldron's Calico Printing Works. This later became a sawmill operated by Locke and Woods [FIG. 82] near Orwell Road. The tailrace turned southwards back to join the Dodder upstream of Orwell Bridge, also known as Waldron's Bridge.

MILL NO. 22 WILLANS' CLOTH MILL, served by a millrace that started at Orwell Weir and flowed into a millpond west of Dartry Cottages [MAP NO. 42 AND FIGS. 83, 84 & 85]. In 1844 the mill was run as a cloth mill. In her book *Thomas Edmonston and the Dublin Laundry*, Mona Hearn notes that between 1845 and 1860 the Dartry Mill was run as a laundry by Thomas Bewley. It was then worked as a paint mill and was in ruins in 1880.

MAP NO. 40
River Dodder Mills 21 to 28
(Mallet's Map, 1844)

MAP NO. 41
River Dodder Mill 21
(OS, 1843)

MAP NO. 42
River Dodder Mills 22 & 23
(OS Map, 1843)

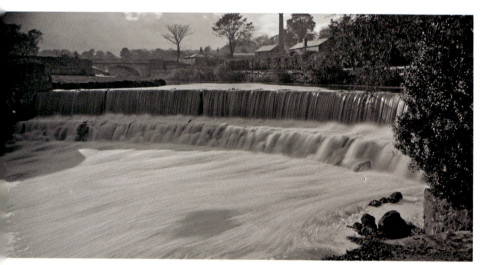

FIG. 81
The weir at Lower Dodder
Road to Mill No. 21
(Robert French, c. 1900)

FIG. 82
Mill 21 – Locke and Woods
Saw Mill with Orwell weir
in foreground

FIG. 83
Weir to Mill 22 at Orwell
Road with Orwell bridge
and Mill 21 in background
(Robert French c. 1900)

FIG. 84
Mill house at Orwell Weir

FIG. 85
Millrace to Mill 22

FIG. 86
Mill 22 - Dartry Dye Works

FIG. 87
Weir to Mill 23 at Dropping
Well, Milltown (Don McEntee)

FIG. 88
Sketch of Dublin Laundry,
Milltown (1888)

This mill was taken over by the Dublin Laundry in 1894 and was fitted out as a dyeing and cleaning works. Hearn stated that "it was felt that there was a need for an up-to-date dyeing and dry-cleaning establishment in Ireland as much of this work was going to England and Scotland". In 1907 the Dartry Dye Works Ltd. took over the dye works branch of the Dublin Laundry [FIG. 86]. The dye works premises were later taken over by North's, a firm of protective clothing manufacturers.

MILL NO. 23 MOORE'S CLOTH MILL, Milltown, upstream of where the Dundrum River joins the Dodder. A millrace flowed from the Dartry weir [MAP NO. 42 AND FIG. 87] and through the grounds of Nullamore, then under the Nine Arches to serve this premises. Up to 1760 the tail race from Mill No. 22 fed the mill pond at Mill 23. The same year, John Classon constructed the Dartry weir to ensure a more secure supply of water to his mills on the site, which was probably the location of the paper mill built by Dupin and leased to Lee, Price and Wild in 1694. As mentioned in the section on Milltown in Chapter 4 there were many other types of mills in the area over the centuries. A number of these were located in the vicinity of the present day Dropping Well where Classon had his sawmills. These were energised from the weir at Dropping Well. In 1844 it was a cloth mill. Mona Hearn notes that the mill on this site had been acquired by Gilbey in 1856 and run as a flour mill in conjunction with a bakery in Patrick Street, Dublin.

Thomas Edmondson, who had run the Manor Mill Laundry in Dundrum with his brother's family from 1863, acquired the site from Gilbey in 1888 and opened the Dublin Laundry [FIG. 88]. It was an ideal site for a laundry with a good supply of water for power and washing clothes and a supply of city gas from the Dublin Gas Company to heat its boilers. It was adjacent to Milltown railway station. The business expanded rapidly, employing over 300 people in 1900. In the early twentieth century two large steam boilers were installed to

FIG. 90
Weir to Mill 24 upstream of
Packhorse Bridge, Milltown
(Don McEntee)

power two Marshall engines which energised the laundry. Following closure of the business on 15th July 1982 [FIG. 89], some numbered piece work pay boxes were retrieved from building as well as two electric vehicles which are in preservation. The millstones are buried in the parkland beside the restored chimney.

A history of brewing in Ireland states that the first lager brewed in Ireland was produced at the Dartry Brewery in 1892; the exact site of his short-lived enterprise has not yet been established.

MILL NO. 24 CLONSKEAGH IRON WORKS. Martin Holland's *"Clonskeagh – A Place in History"* details the history of milling in Clonskeagh.

The first reference to a mill in the vicinity of Clonskeagh is in 1280 when the land was granted to Sir Robert Bagod. The Fleetwood Survey of the Cromwellian period tells us that there was one mill in the townland of Rabuck (on the Dodder at Clonskeagh) "in use worth, in Anno1640, ten pounds". In mid 1700s there were two mills at Clonskeagh: one to the south and one to the north. In the early eighteen century there was a dyestuff factory and a bleach green in Clonskeagh.

The mill race to Mill No. 24 begins at the Milltown Weir [MAP NO. 43 AND FIG. 90]. It flows parallel to the river on the right bank under an arch in Packhorse Bridge and in a channel at the side of the mills on south side of Milltown. The tailraces from the these mills on the Slang River discharged into this mill race. The mill race continued in a culvert beside Milltown Bridge. At Clonskeagh Castle it flowed northwards in an aqueduct entering a large mill pond [FIG. 91] at the ironworks.

Henry Jackson established a large iron mill in 1783 at Clonskeagh. Jackson was a member of the United Irishmen and following the rising of 1798 fled to Philadelphia in the USA. In 1803 he came home but returned to the USA in 1805. Around this time the company was known as Jackson and White. D'Alton in 1838 wrote that "Jackson erected two powerful water-wheels twelve feet in diameter by seven feet wide, one twelve by five and four twelve by three, each possessing a head and fall of sixteen feet with all befitting apparatus of hammers, shears, cylinders and rollers". In 1834 William M'Caskey bought the ironmill. In 1843 it was taken over by John Rochford & Son. For a short period in 1847 he was joined by James Dodd, brassfounder. In 1862 John was succeeded by Henry Hugh Rochford. He expanded the business described as "spade, shovel, scrap, iron shafting, railway and draining tool manufacturer". In 1867 Thomas Henshaw & Co., owners of Mill No. 25, bought this ironworks as a going concern consolidating their business in Clonskeagh. The business continued until 1960. When it closed it was manufacturing spades. The site is now a public park.

In 1894 a legal transaction was recorded whereby the Hibernian Bank became the proprietors of a mill weir and mill race at Clonskeagh. The significance of this is that it reminds us that mill weirs and mill races, being expensive infrastructure essential to mills, were valuable assets.

FIG. 91
Aerial view of Clonskeagh
Ironworks (Mill 24), millpond
and Crampton housing in
Clonskeagh circa 1935
(courtesy of Ruth MacManus)

FIG. 92
"On the Dodder at Clonskea",
painting by J Cunningham
circa 1920 with Clonskeagh
Ironworks on the right
bank of river and workers
cottages on the left

FIG. 93
Weir at Beechhill
Road to Mill 25
(Photo Don McEntee)

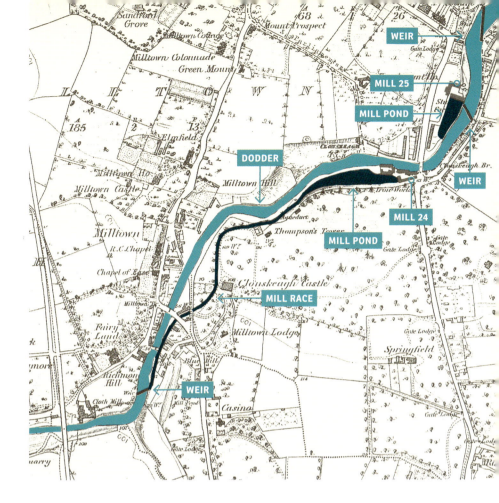

MAP NO. 43

River Dodder Mills 24 & 25

(OS Map, 1843)

J. Cunningham's painting *On the Dodder at Clonskeagh* [FIG. 92] illustrates the iron works on the right bank, labourers' cottages on the left bank and Clonskeagh Bridge in the background.

O'Flanagan, approaching this factory in his late nineteenth century journey along the Dodder, noted "the hum of industry – arising from the clank of hammers, the snort of engines, the whirl of wheels, the din of machinery in full work."

This mill corresponds with No. 9 on the 1879 Smith list where the owner is quoted as the Rev. Dr. McCulloch and the Lessee, Thomas Henshaw.

MILL NO. 25 PORTIS'S IRON MILL. Clonskeagh, 165 metres downstream of Clonskeagh Bridge [MAP NO. 43 AND FIG. 93].

In 1700s there is a record of a mill at this location. Thomas Roberts' (1748-1777) painting "A View of *Clonskeagh*" depicts the mill and mill pond [FIG. 94] at this time. Note the precarious construction of the weir. The *Dublin Evening Post* stated that, on 25 November 1786, the weir of John Carbery's paper mill at Clonskeagh "is entirely swept away by floods". He had just rebuilt the weir two years before the flood. He repaired the weir and continued in business. A drawing dated September 1797 attributed to Brocas shows a paper mill at Clonskeagh [FIG. 95].

In early 1800s Mr Stokes opened an iron mill using three wheels. The next owners were Messrs. Wright and Stanley. The 1837 OS Map shows a stuff factory run by Maiben Baird (1834-1839) and J Baird (1840 -1842). Stuff was a type of coarse thickly woven cloth originally made entirely of wool.

In 1843 Thomas Portis "spade, shovel and edge tool manufacturer" bought the mill. In 1854 it was taken over by Thomas Henshaw. It was associated

FIG. 94
"A View of Clonskeagh", painting by Thomas Roberts (circa 1770) with figures bathing in the mill pond and Mill 25 on right hand side

with the Clonskeagh Iron Works and known as the Lower Mill or Red Mill. Ironmongery for various types of appliances, stamped out at the Clonskeagh Iron Works (Mill No. 24), was finished in this mill. Around 1881 it became a limited company. In 1933 the company was listed as "Clonskeagh Iron Works".

From 1940 the Industrial Appliance and Mill Engineering Co., took over ownership of this mill. In 1953 Jefferson Smurfit and Sons, Ltd., card box manufacturers operated a paper mill here until it closed in 2005.

The mill pond was unique on the Dodder as it was upstream of the Beechill Road Weir [MAP NO. 43] and adjacent to the river.

This mill corresponds with No. 10 on Walsh's 1879 list, where the owner is given as Thomas H. Thompson and the occupier is Thomas Henshaw

BEAVER ROW WEIR AND DONNYBROOK MILL RACE. A long mill race commenced at the Beaver Row weir behind Ashton's pub on Clonskeagh Road [FIGS. 96 & 97]. This mill race dates back to at least the sixteenth century. This weir was formed at a natural rock outcrop in the river. It flowed by Eglinton Park and through the grounds of the Magdalen Convent, where it served the laundry and saw mill [MAPS NOS. 44 & 45]. It then continued northwards flowing parallel with the Dodder north of Donnybrook Green and the eastern boundary of Herbert Park, to Ballsbridge. The watercourse, which today discharges into the Dodder immediately downstream of the bridge at Ballsbridge, formerly continued on

FIG. 95 (ABOVE)
Papermill at Clonskeagh (1797)

MAP NO. 44 (OPPOSITE TOP)
River Dodder Mills 25 to 28
(Mallet's Map 1844)

MAP NO. 45 (OPPOSITE BOTTOM)
River Dodder Mills 26, 27 & 27A
(OS 1843)

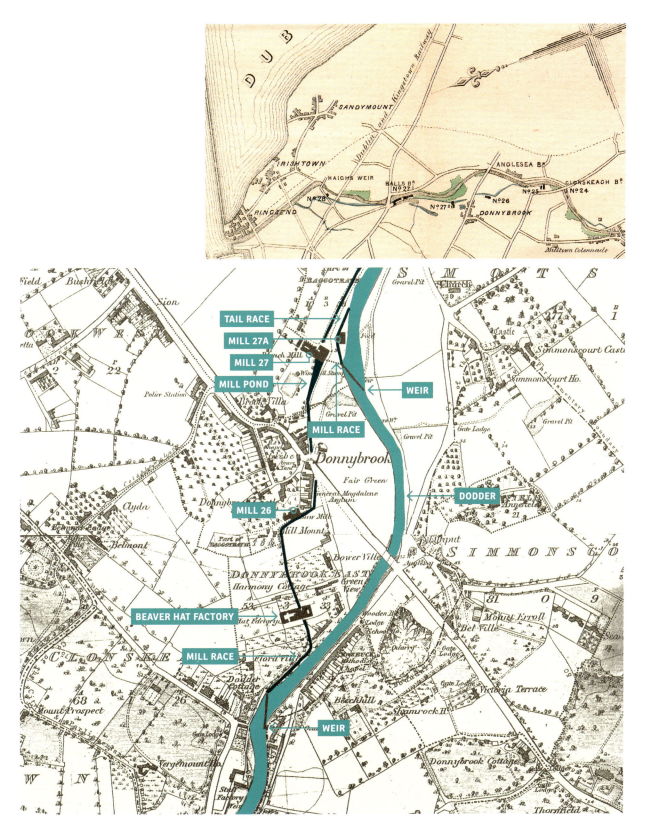

FIG. 96
Clonskeagh Weir,
Beaver Row (Robert
French c. 1900)

FIG. 97

Clonskeagh Weir at Beaver Row

(Don McEntee)

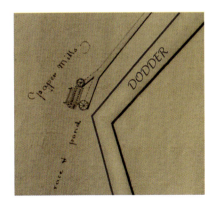

FIG. 98

Paper Mill at Donnybrook (Mill 28) (Samuel Sproule, 1780)

to Mill No. 28 on Mallet's map [MAP NO. 44] in the vicinity of No. 71 Landsdowne Road. It then bifurcated to cross the site of Lansdowne Road Stadium, the two streams joining again to outfall with the Swan River upstream of Londonbridge Road. On Edward Cullen's map of 1731 the mill race is shown flowing to a mill located approximately at the present day New Bridge and continuing on as a mill race joining the Dodder near Irishtown.

Moriarty refers to the mills that stood on the left (north) bank of the river opposite Beaver Row. Joseph, James and Robert Wright built a hat factory circa 1811 where Eglinton Park is today. This was a beaver hat factory where hats were made from felted beaver fur. The workers in the factory were 'felters' from Yorkshire, brought over to make the hats. Beaver hats were fashionable across much of Europe during the period 1550-1850 because the soft yet resilient material could be easily combed to make a variety of hat shapes including the familiar top hat. The Donnybrook Parish Magazine of March 1893 (Vol. IV, No. 93), states that Wright brothers leased part of the lands of Roebuck beside the Dodder in 1811, and built the 20 cottages of Beaver Row for the workers in the hat factory, and a meeting hall behind No. 9 in the Row and a church. The footbridge over the Dodder was constructed for the workers going to the factory [FIG. 26].

There were quarries at each end of the row of cottages. The last quarry in the area was filled in as late as the 1940s and is now the site of Donnybrook No. 2 Bus Garage.

FIG. 99
"Duffy's Cotton Mills,
Ballsbridge" (Mill
28) by James Arthur
O'Connor (circa 1830)

FIG. 100

"Dodder Mill" at Haigs Lane

(Mill 29) by Francis Danby

circa 1812 (Courtesy of Brian Siggins)

MILL NO. 26 A SAW MILL. At Donnybrook Green, which was owned by Hugh McGuirk & Company [MAP NO. 45]

MILL NO. 27 JOHN DUFFY & SON'S BLEACH MILL. Was part of the John Duffy complex of mills in the Ballsbridge area in the nineteenth century [MAP NO. 45]. It was powered by the tailrace of Mill No. 26. Duncan's map of 1821 shows this as a print works and Taylor's map of 1816 shows it as a calico mill.

Mills have been located here in the previous centuries. In 1765 John McMahon and Charles Smith opened two paper mills in Donnybrook. Samuel Sproule's 1780 Pembroke Estate map [FIG. 98] shows a large paper mill between Donnybrook and Ballsbridge powered by two waterwheels driven from a millpond. This was probably one of the paper mills. The second of these may have been powered by a mill race from the Dodder which returned to the river between Donnybrook and Ballsbridge [shown as Mill No. 27A on MAP NO. 45]. The weir is shown on 1843 OS Map but not on subsequent maps.

Archer's 1801 Statistical Survey of County Dublin states that there were two cotton wash mills at Donnybrook owned by a Mr. Dillon. These were probably Mill Nos. 27 and 27A.

MILL NO. 28 JOHN DUFFY & SON'S CALICO & COTTON. Mill located upstream of the bridge at Ballsbridge was powered by the tailrace of Mill No. 27 [MAP NO. 46]. Duffy's mills complex had been built on the site of a sixteenth century mill, one of several in the area owned by Nicholas Duffy, who lived in an old castle here.

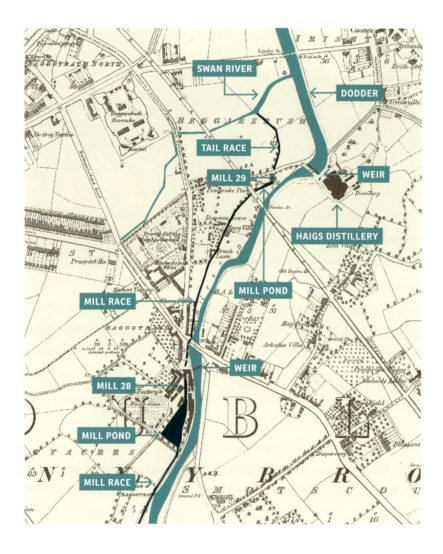

MAP NO. 46
River Dodder Mills 28 & 29
(OS, 1843)

This factory is said to have been established around 1740 at the beginning of cotton manufacture in Europe. Samuel Spoule's 1780 map [MAP NO. 47] locates a mill in Ballsbridge downstream of the bridge. This was probably the first cotton mill in Ballsbridge. Edward Cullen's map of 1731 shows a mill in this same location.

As the demand for cotton grew the cotton mill referred to by Mallet was established upstream of Ballsbridge and over the years expanded eventually employing up to 500 people. The bleach green for Duffy's mills extended along the north bank of the Dodder from Donnybrook to Ballsbridge. Herbert Park now occupies what was once the bleach green. O'Flanagan states that this factory was energised by engines of forty horsepower. James Authur O'Connor's nineteenth century painting [FIG. 99] is a good illustration of the cotton mill. In the mid nineteenth century Duffy's mills were purchased by a syndicate of Manchester firms who closed and dismantled them to crush out Irish competition.

MAP NO. 47

Mills at Ballsbridge (Mill 28) and Haigs Lane [Lansdowne Road] (Mill 29) (Sproule, 1780)

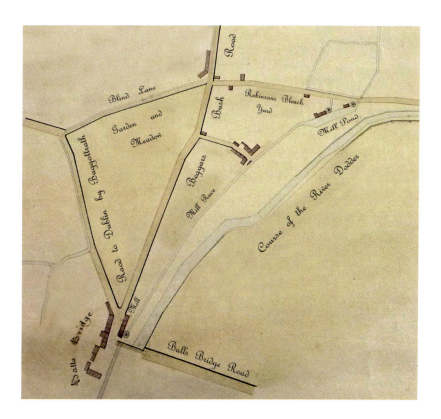

The 1897 Ordnance map shows a Flour Mill at this location with a Bakery beside it. The 1912 OS Map shows Bakery and with a Female Penitentiary to the west side of it.

In time, Burke's Flour Mill at Lansdowne Road moved its operation to Ballsbridge and later became part of the large Johnston, Mooney & O'Brien bakery.

The three firms of Johnston, Mooney and O'Brien existed from early nineteenth century in Dublin. In 1889 the companies merged to form Johnston, Mooney & O'Brien Ltd. It emerged as one of the large industrial bakeries in Dublin who milled their own imported wheat. In 1889 it opened a flour mill and bakery in Ballsbridge. In time the company consolidated its bakery operation in Ballsbridge. This bakery extended many times during its existence. The company's operation included the Clonliffe Mills on Jones's Road, one of the larger Dublin mills, which had been fitted with a steam powered roller system to grind wheat. At the western side of the bakery, one of the Johnston, Mooney & O'Brien's expansions took over a yard previously used by the Pembroke Town Commissioners (site of the former Female Penitentiary). It ceased production at this location in 1989.

MILL NO. 29 BURKE'S FLOUR MILL. , which was also called Donnybrook Flour Mills, and was energised by the tail race from Mill No. 28, was on the most northerly portion of Duffy's holding at Lansdowne Road. Francis Dandy's painting [FIG. 100] shows the mill circa 1812 with two millwheels.

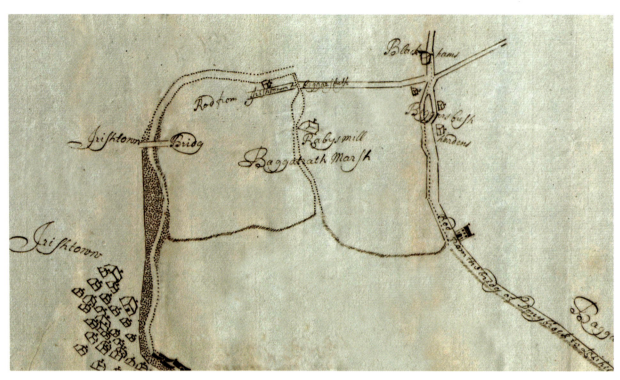

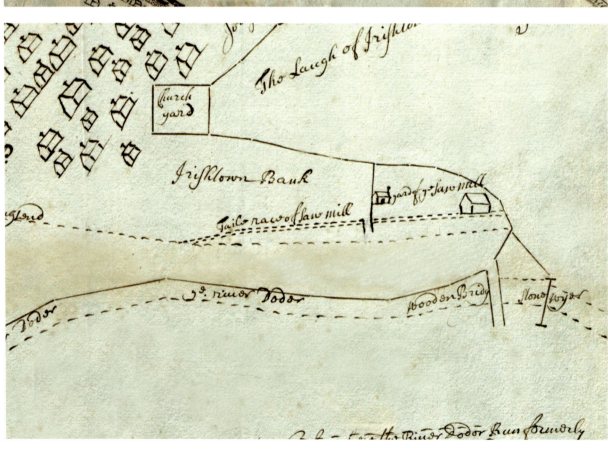

Dodder Bank Mill, described as a Malt Manufactory, occupied the site of the present No. 71 Landsdowne Road.

Samuel Spoule 1780 Map [MAP NO. 47] shows two mills on this site fed from one large mill pond. They were probably cloth mills as they were located on property of Robinson's Bleach Yard.

The disputes that could arise between millers was well described by Lord Fitzwilliam's servant when he wrote on 6th December 1740: "There is an article for a law suite carrying on against one Chappel, who has diverted the millrace leading to the mill at Ballsbridge. I did all in my power to do and persuade the man to settle this matter amicably, but in vaine, he is a wrongheaded fellow and knows not the charge attending these matters. I have hitherto avoided Brangler, but this is an affaire of importance, the subsisting of the mills depending thereon and indeede, they are otherwise in a bad way, for the River Dodder, by last week's inundations, broke the mill race banks at Ballsbridge, made its current thro the road leading to Beggarsbush and so pours on to the Lord Chief Justice Rogerson's ground".

HAIGH'S DISTILLERY [MAP NO. 46]. In the early nineteenth century, the section of Lansdowne Road east of the Dodder was known as Haigh's Lane and Distillery Lane. Until 1833, it led to the notorious Haigh's distillery, which in its later years did not depend on water power — Haigh's Weir apparently being used to pond up water for distillery purposes only. Haighs (not to be confused with the Scottish Haig's) were in a constant state of hostility with the authorities and grim reports, probably apocryphal, insinuate that some over-zealous Revenue inspectors who entered the distillery could not be accounted for subsequently. The distillery was also reputed to have a malthouse on Lansdowne Road (Mill No. 29).

MILLS ON DODDER ESTUARY. J Cullen's 1706 Map [MAP NO. 49] and Edward Cullen's 1731 Map [MAP NO. 11] (both surveyed for Lord Fitzwilliam) show a number of mills along the banks of the Dodder.

J Cullen's 1692 map [MAP NO. 48] shows Rabys Mill on Baggotrath Marsh. This mill could have been powered by the mill race from Ballsbridge or the "Swaney" River which had been diverted at Beggarsbush to form a mill race flowing into the marsh as shown on Edward Cullen's 1731 map.

A premises at Irishtown designated "an old mill" on Sproule's map of 1780 is probably the sawmill shown on J Cullen's 1731 map. Beside the saw mill there was a "*stone wyer*" and wooden bridge [MAP NO. 49].

MAP NO. 48

Rabys Mill, Baggatrath Mash
(James Cullen, 1692)

MAP NO. 49

"Stone Wyer, Saw Mill and Wooden Bridge" at Irishtown
(James Cullen 1706)

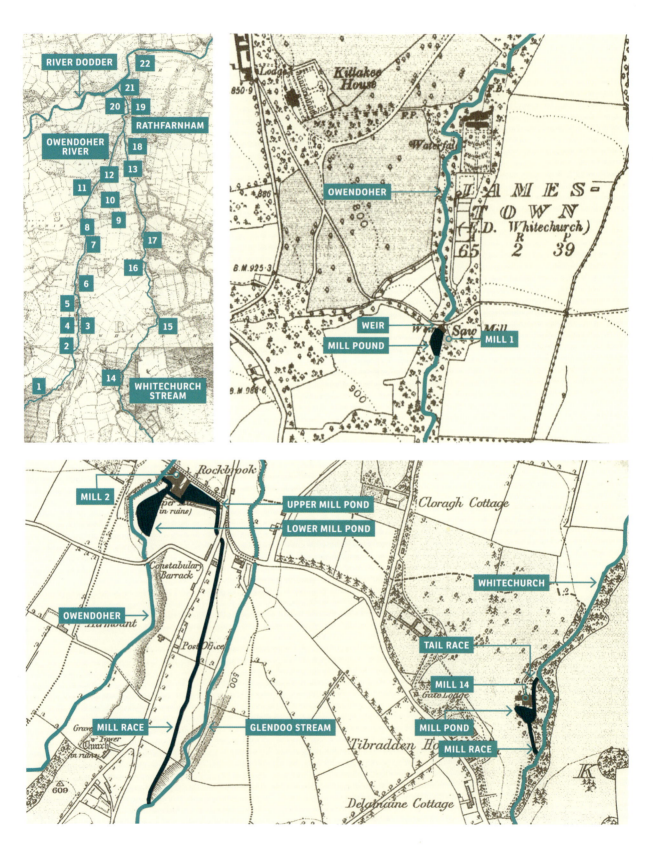

CHAPTER 13

MILLS ON THE OWENDOHER CATCHMENT

The Owendoher is also known as the Rathfarnham River and the Cruagh River, as stated by JR O'Flanagan in a series of articles about the Dodder in *The Dublin Saturday Magazine*, published in the nineteenth century. It is 10km (6.2 miles) long and formed by two streams. The Killakee Stream comes from the Killakee Mountain to the west. The Glendoo Stream comes from the Cruagh Mountain to the east. These branches merge near Rockbrook cemetery and the river then flows north to Ballyboden and on the Rathfarnham.

The Whitechurch Stream, known as the Glin River is a major tributary of the Owendoher. This rises near Tibradden, flows northwards through Grange Golf Course and St. Enda's to join the Owendoher at Willbrook. The Whitechurch Stream is 8km (5 miles) long and has a catchment of 8.3 square kilometres (3.4 sq, miles). Serious flooding occurred on this stream in 2007.

The Owendoher joins the River Dodder south of Bushy Park near the former settlement of Butterfield and just southwest of Rathfarnham Bridge. The Owendoher drains 21.2 sq. km (5,230 acres).

Many paper mills were erected on the Owendoher in the eighteenth century. Twenty two mills are shown on the 1843 Ordnance Survey map and they are numbered as shown on MAP NO. 50. At the beginning of the nineteenth century, as the demand for handmade paper declined, the mill owners changed the use of their premises to cotton and wool production, and later to flour mills. In the early part of the 1800s there were up to 600 people employed in the five mills in the Rockbrook area alone. However, by the middle of the century D'Alton described the same village as one of "squalid paupers" due to the decline and closure of many of the mills.

MAP NO. 50 (OPPOSITE TOP LEFT)
Mills on Owendoher River and Whitechurch Stream, Rathfarnham

MAP NO. 51 (OPPOSITE TOP RIGHT)
Owendoher River Mill 1
(OS 1911)

MAP NO. 52 (OPPOSITE BOTTOM)
Owendoher River Mill 2
& Whitechurch Stream Mill 14
(OS, 1843)

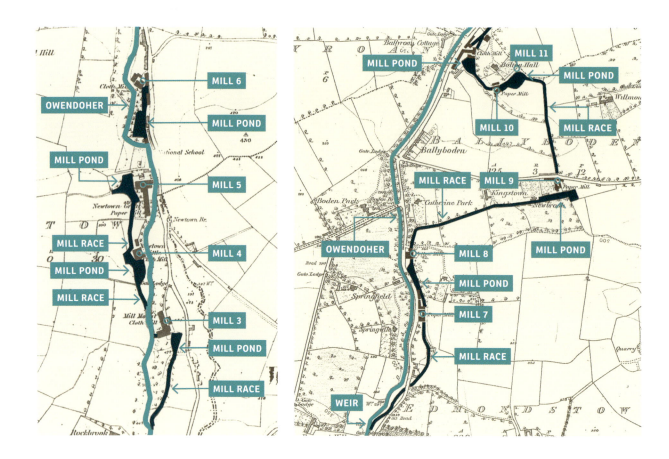

MAP NO. 53 (TOP LEFT)
Owendoher River Mills 3 to 6
(OS, 1843)

MAP NO. 54 (TOP RIGHT)
Owendoher River Mills 7 to 11
(OS, 1843)

The introduction of the steam engine eventually eliminated the need for water power and thus for industries to be located in these otherwise rural locations. A number of the mills were converted into laundries at the end of the nineteenth century, making use of the water in the Owendoher River. A new laundry was opened at Bloomfield near the junction of Ballyboden Road and Whitechurch Road. Details of milling in the catchment are mentioned by Patrick Healy in *Rathfarnham Roads,* in James W Phillips's doctoral thesis *Printing and Bookselling in Dublin 1670-1800* and in Stephen Browne's research dissertation *Success of Employment and Housing in Rathfarnham from 1850 to 1911.*

Due to the steep gradient of the river, this small catchment was able to support the 22 mills shown on the 1843 Ordnance Survey map. Mill No. 1 was on the Owendoher at Jamestown [MAP NO. 51]. The mill race for Mill No. 2 [MAP NO. 52] on Owendoher was fed from the Glendoo Stream, a major tributary of the Owendoher. The mill ponds to Mills No. 3 to 6 [MAP NO. 53] were fed directly from the Owendoher River. The mill race for Mill No. 7 started at a weir on the Owendoher at Edmondstown. The flow from Mill No. 7 tail race then fed the mill ponds at Mills Nos. 8, 9, 10, 11, 12 and 13 [MAPS NOS. 54 & 55] and finally discharged into the Whitechurch Stream. Mill No. 14 [MAP NO. 52] was the first mill on the upper reaches of the Whitechurch Stream. Mills Nos. 15, 16 and 17 were the next on the Whitechurch Stream [MAP NO. 56]. A mill stream from the

FIG. 101

Owendoher River Mill 2,
Rockbrook (Don McEntee)

Whitechurch Stream, known as Landy's Mill Race, fed the mill pond at Mill No.
18, [MAP NO. 55]. It then flowed to the fish pond at Rathfarnham Castle and Mill
No. 19, and on to the mill ponds at Mills Nos. 20, 21 and 22 before discharging
into the Dodder River.

MILL NO. 1 The saw mill at Jamestown [MAP NO. 51] operated from the mid
nineteenth century. A weir was constructed across the Owendoher to form a
mill pond in the river. It operated into the twentieth century.

MILL NO. 2 FRY'S PAPER MILL. The second mill on the Owendoher was the paper mill
of Thomas Watson who began paper manufacture here in 1772. How long the Watson
firm operated the mill is not known. It is listed in the Dublin directory until 1782, the
year Thomas died at his Rockbrook home. In Archer's survey of 1801 a Mr Fry is listed
as the owner. In the 1837 Ordnance Survey map it is listed as a paper mill in ruins.

This paper mill was a major enterprise at the time. There were two large
mills on the site. The mill race from the Glendoo Stream took most of the flow
in the stream down to the upper mill pond [MAP NO. 52]. This pond powered the
mill wheel to upper mill. The tail race from the upper mill flowed into the lower
mill pond which powered the mill wheel to the lower mill. The extensive remains
of these mills still survive beside the avenue to Rockbrook House [FIG. 101]

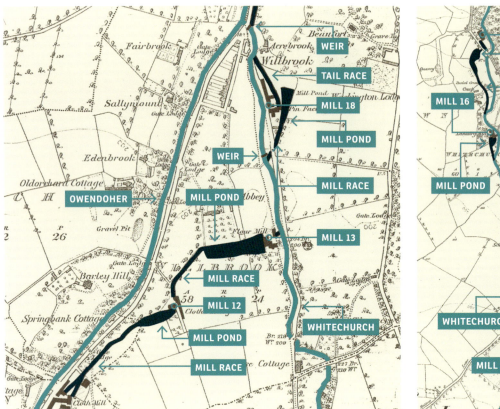

MILL NO. 3 MILLMOUNT MILL. In 1758 Benjamin Nun, a linen draper and Thomas Watson, a bookseller, commenced the manufacture of paper at Millmount [MAP NO. 53]. The partnership dissolved in 1772 when two firms grew out of this one. Nun died in 1776 and was succeeded by his son and partner Richard who continued the manufacture of paper at Millmount until 1794. In that year he was joined by Clement Taylor and Nicholas Graham, two Englishmen, to take advantage of the duty imposed on imported paper from which the Irish papermakers benefited. Taylor was an important manufacturer of paper in England.

The expensive equipment installed in the mill to bring it up to the standard of the best in Europe involved the firm in debt. In 1799 Richard Nun unsuccessfully petitioned the Irish House of Commons for a grant to rescue the company from its difficulties; it was bankrupt in 1800. It is mentioned by the same name in Archer's survey of 1801 but later passed to Messrs. Dollards who operated it in conjunction with their paper mill lower down the river. Here the rags were processed and made into the raw material from which the paper was manufactured. This mill closed in 1899.

MILL NO. 4 NEWTON LITTLE CLOTH MILL. This may originally have been a paper mill. It was converted into a cloth mill before 1836 and seems to have ceased production shortly afterwards.

MAP NO. 55 (TOP LEFT)
Owendoher River Mills 12 & 13 and Whitechurch Stream Mill 18 (OS, 1843)

MAP NO. 56 (TOP RIGHT)
Whitechurch Stream Mills 15, 16 & 17 (OS, 1843)

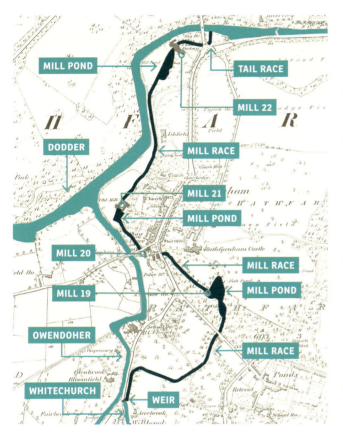

MAP NO. 57
Whitechurch Stream Mills
19 to 22 (OS, 1843)

MILL NO. 5 NEWTON GREAT PAPER MILL. Robert Randall served his apprenticeship to Edward Waters at Waters's mill in Milltown. In 1729 he leased from Thomas Talbot the mill at Newbridge near Leixlip, Co. Kildare for a term of 31 years and converted it into one suitable for making paper. He continued to operate the Newbridge mill until 1750, simultaneously expanding his business. With a grant of £300 he built this paper mill in 1757 near Little Newton as well as owning a white paper mill near Rathfarnham. He advertised the Newton mill for sale in 1759. His obituary notice in 1781 reveals that he was "formerly an eminent papermaker" indicated that he had retired.

In the early 1800s Mr Pickering of the London House of Magnay Pickering operated this mill. He was working it in 1836 but some years it was worked by Mr Brown who closed it down about 1866 when he went to America. A school now stands on the site of Newton Great Paper Mill.

MILL NO. 6 WOOLLEN MILL LATER KNOWN AS RECKITTS. Reckitt's well-known factory is on the site of a woollen mill which was operated by John Reid of Ballyboden until 1892 and by Frederick Clayton & Sons until 1909.

MILL NO. 7 EDMONDSTOWN PAPER COMPANY. Beside the row of mill workers' cottages there was a paper mill [MAP NO. 54] operated by Messrs. Dollards from 1848 until 1896. It was subsequently operated by the Edmonstown Paper Company until about 1912.

MILL NO. 8 SHERLOCK'S COTTON MILL. The tail race from the paper mill fed this cotton mill. Owned by Robert Sherlock, it was one of the smallest local mills in the mid-nineteenth century but still employed about 20 people. It was converted into the Edmondstown Model Laundry Company in 1873 and initially run by Sherlock. By 1901 Richard Hughes was operating the laundry, which closed down before 1920.

MILL NO. 9 NEWBROOK MILL. The tail race from the cotton mill discharged to the mill ponds at Newbrook Mill on Taylor's Lane. Extensive paper making was carried out here for many years. Mr John Mansergh who started the business died in 1763. John McDonough ran the mill from 1846 to 1897. The two nearby residences, Newbrook House and Kingston House, were usually occupied by those engaged in the industry. John Hughes lived in Kingston House and his relations Richard and Thomas McDonough lived in Newbrook House. From 1901 to 1935 the industry was operated by Sir John Irwin who during this time lived at Newbrook House. The mill was extensively damaged by fire in 1942 and has since been demolished.

MILL NO. 10 RYAN PAPER MILL. The flow from Newbrook Mill passed to the paper mill and then to Reid's Woollen Mill beside Bolton Hall. Behind Bolton Hall there was a paper mill belonging to Nicholas Ryan which is shown on the map of 1837. The owner's name was changed to Simon Brown in 1854 and to Joseph P Brown in 1863. It ceased to operate from 1876.

MILL NO. 11 REID'S WOOLLEN MILL. At the northern end of Ballyboden village there was a woollen cloth mill which belonged to John Reid from at least as far back as 1836 when D'Alton in his history of Co. Dublin states that this mill gave employment to about 40 people. It was owned in 1874 by John and James Reid. The valuation was reduced because of disuse and in 1874-1896 it was recorded as a woollen cloth mill (idle). The ruins of this mill were demolished in 1950 when the entire village of mill workers' cottages was cleared and a brand new village of council houses was erected on the site.

MILL NO. 12 MILLBROOK MILL. On the opposite of the road to Orchardstown Estate stood Millbrook Mills otherwise known as the Little Mill. This was shown as a cloth factory on the 1837 OS Map [MAP NO. 55]. The diameter of its water wheel was sixteen feet. By 1848 it was being operated as a flour mill by Michael Nickson. In 1881 RH Fenwick was working it as an iron mill but Richard Dempsey changed it back to a flour mill in 1883. It ceased operating in 1887.

MILL NO. 13 WILLOWBROOK FLOUR MILL. St Gatien's, on the western side of the Whitechurch Road, occupies the site of Willbrook Flour Mills. Noted by D'Alton as Egan's Flour Mills in 1836, it was held by the Gibneys from 1847 to 1872 and by John Garvan from 1873 until 1882. After a few years of lying idle it was incorporated with a dwelling house by JE Madden in 1885 and named St. Gatien's.

MILL NO. 14 SILK MILL. The first mill on the Whitechurch Stream, shown on the 1843 OS Map [MAP NO. 52] was an Old Silk Manufactory at Tibradden owned by Mr. Hughes

MILL NO. 15 WOOLLEN MILL. a woollen mill at St. Thomas [MAP NO. 56]. Woollen cloth was manufactured here by Thomas Thorncliff until about 1880. Previous to this Ernie Shephard mentioned that this was a corn mill owned by Dolan.

MILL NO. 16 LAUNDRY MILL. This laundry was started around 1830 by Thomas Bewley in whose family it remained until 1880. It was then operated by Caroline Thacker and from 1899 by Mr. Willoughby. It closed down around 1930.

MILL NO. 17 SILK MILL. In 1756 William Mondett and Moses Verney came from England and built a paper mill at Whitechurch at a cost of £1,400. By 1763 they had built a second mill at the same place to make both press and purple papers. The laundry mill on the other side of the road may possibly have been the location of the second mill. Verney worked the mills until 1786. This mill later became Jackson's Cotton Mill, of which extensive remains still existed in 1836. On the 1837 Ordnance Survey Map it was shown as a silk mill.

MILL NO. 18 SILVERACRES MILL. This was named Brooklawn Mill on Taylor's Map of 1816 and on Duncan's of 1821. In 1836 Mark Flower had a pin and wire factory here which was then named Silveracres Mill [MAP NO. 55]. It closed in 1853 and was converted into a flour mill by Robert Gibney who also owned the nearby Willbrook Mills. It was operated by Patrick Gibney from 1864 to 1893, after when it was taken over by JE Madden. Subsequent to 1899 it changed hands frequently and the last tenant was Mr Murray from 1922 to 1933. The mill has since been demolished but the Mill House and some out-offices still remain.

MILL NO. 19 RATHFARNHAM CASTLE SAW MILL. After passing through Silveracres Mill the flow entered a mill race which crossed Nutgrove Avenue into the grounds of Rathfarnham castle and fed several fish ponds. A saw mill was located at the exit from the fish ponds [MAP. NO. 57].

MILL NO. 20 SWEETMAN'S FLOUR MILL. From the fish ponds in Rathfarnham Castle the flow was conducted under the road to the flour mills which stood at the corner of Butterfield Lane. Described in 1836 as Sweetman's Flour Mills, it employed up to fifty people and frequently changed hands before closing down in 1887. The flow then crossed to the north side of present day Butterfield Avenue.

MILL NO. 21 RATHFARNHAM PAPER MILL. Church Lane leads to Woodview Cottages, which were built partly on the site of the old paper mill. The mill race from Mill No. 20 passed under Butterfield Lane to the paper mill and continued on below Ashfield to turn the wheel of the Ely Cloth Factory. The paper mill has been described as the oldest in Ireland.

Lord Wharton granted a lease to Dupin's Company of White Papermakers to construct the mill to produce paper. This was producing paper in 1693 with Dupin's watermark. In 1697 the firm, then under the direction of Colonel John Perry, petitioned the Irish House of Commons for an extension of the time period granted by the letters patent. This was not granted and the company dissolved sometime after 1897, probably in 1701.

In 1701, Lord and Lady Wharton leased the mill to Thomas Jones. In 1717, after Thomas Jones died, William Lake, who was Jones's son-in-law, took over ownership of the mill. He unsuccessfully petitioned the Irish House of Commons for monetary aid in the pursuit of papermaking in 1719. Jones transferred the mill property to John Foules of Rathfarnham in 1720.

The year 1726 marked the first outright entry of the Slator family into the field of papermaking in Ireland when Joseph Slator acquired this mill. Slator leased the mill from William Connolly, to whom the lease reverted in 1724, the terminal date specified in the lease to Thomas Jones. In 1742 a new lease was drawn up, the deeds being transferred by Joseph to his sons Thomas and William in the following year.

Thomas Slator died in 1763 and his obituary noted that he had worked in the Rathfarnham mill since he was eleven years of age. One might deduce from this that his father before him was a papermaker — perhaps one of those imported by the Company of White Papermakers when they came to Ireland. The works were destroyed by fire in 1775 and rebuilt the following year. Following William's death in 1768, Thomas continued to operate the mill until he, too, died in 1776. His son Joseph kept the mill in operation until 1792.

Randal, who had a paper mill at Newbridge as well as the Newton Great Paper Mill, also had a white paper mill near Rathfarnham. This mill may have been constructed beside Slator's as Archer's survey of 1801 mentions two paper mills here, Freeman's and Teeling's. Both Dalton in 1836 and Lewis in 1837 state that one paper mill was still working and from 1836 to 1839 the operator's name was Henry Hayes. Rathfarnham Mill appears in the directories and if this can be identified with the mill at Woodview cottages it must have become idle soon afterwards as it was designated "Old Mill" on the 1843 edition of the OS Map. In 1854 when this mill had neither water wheel nor machinery an attempt that was made to re-open it for the manufacture of paper came to nothing

In the 1860s JR O'Flanagan wrote in the *Dublin Saturday Magazine:* "near Rathfarnham stood an old mill, where some two centuries ago paper was first made in Ireland. It was on the produce of this mill, we are informed, Usher's "Primordia" was printed, and also, what must be interesting to every lover of Irish literature, it was on paper produced there the *Annals of the Four Masters* was written. There may be an error here as the Annals was produced before this mill was established and the version of the Annals referred to by O'Flanagan was probably a new edition.

MILL NO. 22 ELY CLOTH FACTORY. A millpond and extensive mill buildings formerly occupied the low-lying fields on the west side of the Rathfarnham Road, just beside the bridge. On a map by Frizell dated 1779 it is called the Widow Clifford's mill and mill holding and on the 1843 OS Map it is called the Ely Cloth Factory. The Ely Mills comprised a woollen mill and a factory. According to Lewis, at the time of his survey in 1831 the mill employed approximately 100 people and the owner was a Mr Murray. In 1850, it passed into the hands of Mr Nickson who converted it into a flour mill. His family continued in occupation until 1875 when John Lennox took over. The mill closed in 1880.

CHAPTER 14

MILLS ON THE SLANG (DUNDRUM) RIVER

The Fitzwilliams were living in Merrion Castle, when, in 1593, Richard inherited the estate in Dundrum and decided to build a new castle , on the ruins of the 13th century castle abandoned over a century before. He died only three years later and his son, Thomas, was the first resident in the castle when it was completed. Many Inquisitions were taken in those times and those for Leinster were published in 1826. Among them are two relating to the possessions of Thomas in Dundrum. The first, taken in 1627, lists 1 castle, 6 dwellings (*mes'*) and 200 acres; the second, in 1639, adds "*1 molendin' aquatic'*", a water mill. This is the first reference to a mill in Dundrum. It is possible that the monastery at St. Nahi's had a mill on the river over a thousand years ago but no evidence has been found.

As a result of the 1641 rebellion, a Civil Survey was carried out in 1654, to ascertain the ownership of land and the assets thereon. No mills were recorded on the River Slang. Dundrum had been held by the rebels for several months and they may have destroyed the mill.

The next reference to a mill on the Slang is in a 1737 deed memorial, between Patrick Reynolds and Nicholas Rogan, recorded by Hogg. Other deed memorials on mills in Dundrum he recorded were: 1746, between Fitzwilliam and Isaac Dobson; 1753, between Samuel Taylor and William Thomas; 1754, between Isaac Dobson and Thomas Reynolds. In "A list of Tennants Names who is living on the lands of Dundrum with an account of what each pays for their Holdings" from the 1740s, 'Crummill the Miller' was listed for £30 (Pembroke Estate Papers).

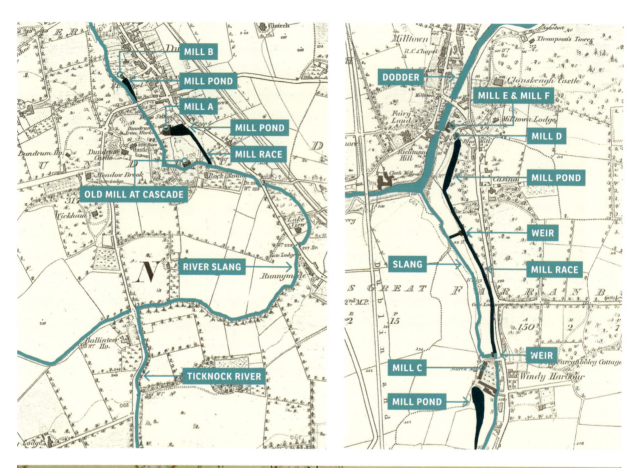

Isaac Dobson was the grandson of a Cromwellian officer granted Dundrum Castle in 1655, subject to the rights of the Fitzwilliams. In 1754, he was accused by Fitzwilliam's agent of diverting water from the mill (Pembroke Estate Papers). There was a cascade in the river near Dundrum Castle, which was then being used for a mill [MAPS NOS. 58 & 59], and a headrace was later taken off above it to feed a mill-pond which became the main source of power for the mill in Dundrum. Dobson, who died suddenly the day after the accusation, may have started constructing the headrace and the other mill owner objected. Jonathan Barker's 1762 map of the Fitzwilliam estate at Dundrum shows a building beside the cascade labelled "mill" but no millpond. A contemporary map by Rocque doesn't shows any mill in Dundrum but shows a paper mill in Windy Arbour, the earliest reference to a mill there. MAPS NOS. 50 & 60 show the locations of the mills on the Slang. The principal establishments are described below.

MILL A. Archer lists two mills in Dundrum in his 1801 Survey. Mr Stokes had an iron works with three mill wheels and Mrs Hall had a paper mill [MAP NO. 58]. Dundrum Iron Works, then owned by George Sikes, continued until about 1850. The mill operated for 10 hours a day for eight months in the year with a head of nine feet. It had two overshot wheels, 12 feet in diameter. The bellows wheel was three feet wide and the hammer wheel four and a half feet, each with about 30 buckets (Hogg).

Hewetson Edmondston started the Manor Mill Laundry on this site in 1863. He brought the very latest laundry technology to this country, using water power to drive the machinery and steam for washing and drying. When the weather was good, the laundry was hung on outdoor lines to dry. Following Hewetson's death in 1871, the laundry was run until 1887 by his brother Thomas and widow Huldah. Her son George then took over the operation, which was continued by the Edmondson family until its closure in 1942. The mill pond was later used to generate electricity for the laundry, which was the largest employer in the area.

When the laundry closed it became the Pye Works, making radios and, later, televisions. The Dundrum shopping centre now occupies the site. Part of the old mill pond has been preserved.

MILL B. Neither Taylor nor Duncan names a mill in Dundrum but Duncan does show a headrace to a mill pond and two buildings across the tailrace, which doesn't rejoin the river until after the bridge on the Ballinteer Road. The first Ordnance Survey Map of 1843 has what appears to be a mill pond at this location behind the church. It may have been Mrs Hall's paper mill but it is more likely to have been the Windy Arbour Mill, which Archer included in Dundrum.

MILL C. This mill [MAP. NO. 60] is shown on Taylor's (1816) and Duncan's (1821) maps as a paper mill. In 1837, according to Lewis, it was 'a silk-throwing factory belonging to Mr (John) Sweeny, Jr, employing about 80 persons' The silk industry in Ireland had been in serious decline following the withdrawal of protection duties in 1826 In 1838, D'Alton noted "At Windy Arbour were the silk works of Mr Sweeny, now deserted." By 1843 (OS Map) it had become a starch mill and

MAP NO. 58 (OPPOSITE TOP LEFT)
River Slang Mills A & Mill B
(OS, 1843)

MAP NO. 59 (OPPOSITE BOTTOM)
"Old Mill" on the River Slang
beside Dundrum Castle
(Johathan, Barker, 1762)

MAP NO. 60 (OPPOSITE TOP RIGHT)
River Slang Mills C, D, E & F
(OS, 1843)

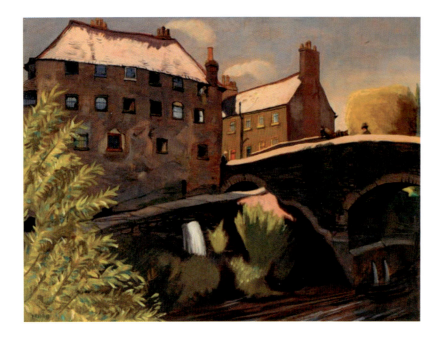

FIG. 102
The Old Millhouse, Milltown
(Harry Kernoff, 1936)

Griffiths records it as a vacant bark mill in 1849. Richard Charles was the lessor of 10 acres which included the mill and nine houses.

A new lease, for 999 years, was made in 1850 and held by William O'Neill, Elm Park, Whitechurch. After his death, the lease was advertised for auction in November 1872. It included the mill pond, a respectable dwelling house and 40 cottages. All the cottages were let, as was the land, but the house and mill were not although there was "a large reservoir and plentiful supply of water in connection with the very valuable mill." There is no evidence of the mill being used again after 1849.

On the right bank of the Slang adjacent to Dundrum Road / St. Columbanus' Road junction at Ryans Arbour House there appears to be the remains of a weir at the side of the river and a millrace coming through an old stone wall which was part of an old building. This would suggest that at one time there was a mill at this location.

MILL D The Slang was also used in Milltown. All mills on the east bank of the Dodder that used the Slang were in Farranboley Townland [MAP NO. 60]. The photograph of circa 1880 [FIG. 20] shows these mills in the complex of buildings beside the Dodder. The mills of Milltown that used the Dodder were on the north side of the river to the west of Milltown. A large mill pond was built in the 1820s on the ridge, later called Millmount, overlooking the village between the river and Dundrum Road, with a long millrace from Windy Arbour. Lewis tells that there was a starch and glue mill here in 1837, shown as a "Blue Mill" on the 1843 OS Map.

In Griffiths Valuation two millers are noted on the Slang in Milltown, the second being James Henderson who had a flour and saw mills, Mill D. He had 12 acres in his holding so he must have had the main mill and possibly shared the mill-pond with John Lee (Mill E), who had no land, between Mill D and the Dodder. The tailrace from Henderson's mill turned the undershot wheel in Lee's mill.

Henderson had a row of seven cottages for his workers, situated near the old bridge. His mill was taken over by Thomas James O'Callaghan in 1855 and was a corn mill for twenty years. He lived in a house beside the pond, appropriately named Millmount.

MILL E. The 1843 Post Office Directory lists William Lee as a millwright in Milltown and Thom's 1846 Directory has John Lee as proprietor of a marble works, although in the notes on the "dilapidated village" the starch and glue mills were still in operation. There is no reference to a mill in the following year's notes. Griffith's Valuation lists Lee's mill as "sawing" and Thom quotes it as "marble and saw works".

MILL F. Another mill came into operation in 1862 when Robert Gibney started a flour mill beside John Lee, sharing the water from O'Callaghan's mill. As there was not sufficient water flowing from the tail race of O'Callaghan's mill to run the two Mills E & F they shared the water. The marble works operated for one week in three and the flour mill operated for the other two. These mills are illustrated on Harry Kernoff painting of *"The Old Millhouse"* [FIG. 102].

Gibney already had two mills in the Rathfarnham area. John Lee died in 1873 and the marble sawing mill closed. T.J. O'Callaghan's mill ceased about 1875 and Gibney's in about 1885. This ended milling on the Slang River. Gibney had a "Millshop and Bakery" in Bankside Cottages for many years after.

CHAPTER 15

A BRIEF STEP FORWARD IN TIME

number of important enterprises were established along the Dodder after Mallet's time. East of Ballsbridge, the strip of land between Shelbourne Road and the Dodder became the location of two major establishments, both now history, that are worthy of note.

Ballsbridge Motors stands on the site of the Dublin Southern District Tramways Company (DSDT) horse tramway depot, opened in 1879. During the years 1892-1895 a steam-powered generating station was built here and the depot was reconstructed to accommodate electric cars. Dublin's first electric trams began operating out of this depot on the Dalkey line on 16th May 1896. The DSDT became part of the Dublin United Tramways system in September 1896.

The Ballsbridge generating station was later superseded by a much larger installation at Ringsend in 1898 but Ballsbridge (Shelbourne Road) depot continued in use as a running shed where major tramcar overhauls and rebuilds were also carried out. It closed following the withdrawal of the last trams in 1949 and less than a year later became an assembly plant for Volkswagen cars and vans. Ballsbridge Motors are now on the site.

In 1912 a laundry was established on a site immediately north of the tramway premises. The laundry adopted the Swastika as its title and trademark, a symbol which at that time was regarded as a token of good fortune. Such was the reputation of the company that the association with the Swastika did it little harm in the 1930s and 1940s. When domestic washing machines, laundrettes and washroom linen replacement services caused the demise of several traditional laundries in the third quarter of the twentieth century, the Swastika Laundry metamorphosed into Irish Linen Services, but moved away from Ballsbridge around 1990.

The tramway power station and the Swastika Laundry both used Dodder water and discharged waste water back to the river. These discharges, particularly those from the laundry, would today be of concern to pollution control officers. Construction of the tramway depot eliminated the mill race that previously reached the Malt Factory at 71 Landsdowne Road.

Bygone industries of Ringsend include rope making, ship and boat building, lime and salt works, at least two foundries and glass bottle manufacturer. In the early years of the twentieth century, when electricity usage was on the threshold of continuous expansion, there were three major generating stations in or adjacent to Ringsend. Pembroke Urban District Council had its electricity works on South Lotts Road.

The Dublin United Tramways Company opened its generating station on Ringsend Road in 1898 and Ringsend, never served by horse trams, enjoyed an electric service from 1900 to 1940. Dublin Corporation's power plant was at the Pigeon House Fort, which has a long, involved and often related history. In 1927, electricity generation became the responsibility of the ESB, which took over the Corporation's Works and developed the Poolbeg Station in the 1970s.

From 1906, Dublin Corporation had its main sewage pumping station at Ringsend, and an outfall works at Poolbeg. The City Council's new Ringsend pumping station and Poolbeg treatment works came into operation in stages during the 1980s. At that time, the Rathmines and Pembroke sewage outfall, which had been at the Half Moon on the South Wall since 1882 was diverted into the new pumping station.

Electricity generation and drainage treatment continue to provide employment in the Ringsend area.

MAINTAINING CONTINUOUS WATER POWER

Oscillating between inadequate and overwhelming, the flow of water in the Dodder was a constant problem for the mill owners, who often had to curtail or stop work in dry weather. A further worry was the inadequacy of supply when millers upstream closed sluices to create a head of water, cutting off the supply to those below them for as long as the sluices remained closed. Robert Mallet explained how "every miller is dependent on the one above him, who throws him idle the moment he shuts down his sluice to accumulate a head, after exhausting his own temporary supply. Hence delay and disappointment in the execution of orders, and consequent loss of trade, unproductive payment of rents, and charges, and hands thrown idle, and either paid for time by the humane employer, for which he receives no return, or left to their own scanty resources, and this their means of subsistence placed literally at the mercy of the elements."

At the time of Mallet's 1843 survey, three of the 27 mills he inspected were disused, another three had already installed steam engines and two others were considering steam power. Forty years later, Bohernabreena, with its compensation reservoir, would alleviate the misfortunes of the millers as well as the severity and extent of flooding downstream.

In *Glenasmole Roads*, Patrick Healy states that when the reservoirs at Bohernabreena were under construction, Dodder water energised 45 mills, some of which were on the Poddle and City Watercourse. In addition to providing Rathmines with water, the reservoirs also ensured a constant supply to the mills, which in the 1880s included "15 flour mills, the remainder consisting of paper, paint, cardboard, cotton, saw, glue and dye mills, as well as distilleries, breweries, malt houses, foundries, tanneries and a bacon curing factory." It must be pointed out, however, that several of these enterprises were along the banks of the Poddle.

Robert Mallet observed that lake catchments often assisted in the prevention of flooding in waterways. He therefore recommended the construction of a reservoir at Glenasmole, and adduced much scientific and engineering knowledge to prove the efficacy of his proposals for the benefit of the millers. A reservoir behind a dam from St. Anne's Friary on the eastern side of the valley to Glassavullaun on the western side would have been 1,000 feet (305m) long and over 100 feet (30m) high. The proposed surface water area would have been 142 acres and the capacity of the reservoir would have been 227 million cubic feet.

Mallet calculated the cost of all the proposed work at £30,000, but beyond some minor improvements to millraces and the construction of a deep channel downstream of Spawell, no further action appears to have been taken in the mid-1800s. If the possibility of using the Dodder as a source of pure drinking water was considered at that time, it does not appear in the Mallet report.

VANISHED ENTERPRISES

In 1991 CL Sweeney wrote "The Dodder, for countless ages, down to the end of the last (nineteenth) century turned mill wheels all along its course. Corn mills, Cloth mills, Flour mills, Tuck mills, Saw mills, Paper mills, Iron mills, Calico print factors, Laundries and Dye Works were operated by its millraces.

"All these millraces are now abandoned and mostly dry, covered in, erased and forgotten by the grandchildren of yesteryears generation." In the twenty-first century, they are a prime subject for study by historians and industrial archaeologists. Among the many victims of change in the mid-nineteenth century were flour mills, which declined in numbers following changes in the Corn Laws and the growth of companies that combined milling and baking on an industrial scale.

With the demolition of the mills, their water wheels and other features there disappeared whole swathes of technology, craftsmanship and lifestyles. Industrial archaeologists have to work very hard to ensure the conservation and restoration of what little remains. Extant or defunct, a handful of workplaces along the Dodder and especially the people who worked in them, are recalled by vehicles in the Transport Museum donated from the Dartry and Swastika Laundries, the Johnston Mooney & O'Brien Bakery, Donnybrook and Ballsbridge bus and tram depots.

CHAPTER 16

RATHMINES WATER SUPPLY FROM THE GRAND CANAL AND THE DODDER AT BOHERNABREENA

n August 1860—sixteen years after Robert Mallet's Dodder report—a Royal Commissioner, Sir John Hawkshaw, investigated Dublin's water supply. He examined six different proposals to replace the polluted and inadequate canal supplies from commercial entities interested mainly in profit. It is noteworthy that he took less than two months to produce his report.

Although it was the most expensive of the six schemes examined, Sir John Hawkshaw recommended the Vartry proposal for two important reasons. More than adequate to supply the City, it could also serve Wicklow, Greystones and Bray, together with all the urban areas and townships along or near its route from Roundwood to Dublin. Starting with Kingstown in 1834, existing legislation and a series of Improvement Acts created nine independent townships in the Dublin area up to 1878. Five of these shared a boundary with the City, which in the case of Rathmines was the Grand Canal. The townships offered comfortable housing in attractive locations away from the decay and stench of the city.

Rathmines Township, established in 1847, had three subsequent boundary extensions; the first was in 1862, when it was renamed Rathmines and Rathgar. The Harold's Cross area became part of Rathmines & Rathgar in 1866, followed by Milltown in 1880. The Milltown enlargement resulted in the Dodder becoming the township's southern boundary. The population of Rathmines and Rathgar, only about 10,000 in 1847, increased steadily to a maximum of about 30,000 in the 1890s. The township pursued a policy of hostility towards Dublin Corporation and was the only one of the nine eligible local authorities not to take Vartry water, all negotiations while the scheme was under construction failing on the issue of cost and quantity.

MAP NO. 61

Map of Bohernabreena
Reservoirs circa 1935
(OS Map surveyed 1908
reprinted 1933)

181

Work on the Vartry scheme began in 1861 and when the system came into operation in 1867, it powerfully enhanced the quality of life in the areas it served. Meanwhile, and astonishingly in the light of Dublin's unhappy experience with canal water, Rathmines and Rathgar established a compact waterworks on the Grand Canal at Gallanstown, near Clondalkin. While more sophisticated than Dublin Corporation's erstwhile canal supplies, this system was, as predicted, a deficient and unsatisfactory enterprise even as it came into service on 23rd July 1863.

The success of the Vartry scheme and the fact that their less affluent neighbours in Dublin enjoyed an abundant supply of pure soft water from 1867 onwards greatly unsettled the residents of Rathmines. By 1877, their ill-conceived and increasingly troublesome water system was so inadequate that the Rathmines Commissioners were embarrassed into looking for something better than the insufficient supply of inferior quality coming from the Grand Canal. Pressure was low and a pumping station had to be installed at Harold's Cross, together with a water tower at Rathgar, to serve the higher areas of the township. The already serious situation became even more intolerable during the very dry summer of 1877.

Pertinent facts and figures were quoted in a letter in the *Daily Express* on 19th July from Frederick Stokes, Chairman of the Rathmines and Rathgar Commissioners. The average daily water usage of each Rathmines resident was 50 gallons (227 litres), the total for the then population of between twenty and twenty-two thousand coming to a million gallons per day. Stokes reckoned that 30 gallons (136 litres) a head should suffice daily, including road watering. He pointed out that the other townships received 20 gallons (90 litres) per head each day from Dublin Corporation's Vartry scheme. The chairman concluded by making the point that so much water was being wasted during the night that as soon as the watering carts set to work in the morning, the supply to houses on the highest levels was affected.

After negotiating once again with Dublin Corporation for a Vartry supply but (predictably) disagreeing on cost, the Rathmines Commissioners finally decided to have their own up to date and independent water supply system. In 1887 they engaged the experienced engineer Richard Hassard, M.Inst.C.E. who had first suggested the Vartry scheme to Dublin Corporation. Hassard was to advise the Commissioners as to the feasibility of obtaining additional water from the Grand Canal or an independent supply from the Corporation reservoir at Stillorgan.

When presenting his report, Hassard also laid before the commissioners, on 1st September 1877, a project by which a supply of water might be obtained by gravitation from the tributaries of the River Dodder draining into the Glenasmole Valley. On 17th September 1879, this proposal was accepted by the Commissioners, who then promoted a Bill in Parliament to obtain the necessary powers for the construction and operation of a system on the River Dodder that would have a three-fold purpose.

Richard Hassard was not the first to propose the Dodder as the source for a municipal water system. In 1854 Parke Neville, the City Surveyor (and City

Engineer from 1857) had contemplated an impounding reservoir at Glenasmole and a service reservoir at Kimmage. Situated three miles from the city centre, the Kimmage reservoir would have been at a level of 170 feet (52m) over datum and have a capacity of 380 million gallons (1,363 million litres).

In addition to supplying the city at a reasonable pressure, it would also have been capable of meeting the needs of the Rathmines Township, then a mere seven years in existence. There would, however, be a problem with water from the peaty upper reaches of the catchment, from which the colour could not be removed using contemporary technology.

Robert Mallet had suggested the Dodder to the 1860 Hawkshaw inquiry but his proposal was turned down because, among other reasons, the quantity it could provide would probably be insufficient to meet Dublin's needs. But mindsets that prevailed twenty years later would greatly benefit water consumers in Rathmines, ease the uncertainties of millers and provide a modicum of flood protection to riparian properties along the Dodder. Mallet, who died in 1881, would surely have been pleased.

The location then chosen to impound Dodder water for Rathmines was at Glenasmole — or Bohernabreena, the name by which the complex is known today. Dublin Corporation unsuccessfully opposed the project, most fortunately in the light of subsequent developments. Research by Michael Murphy reveals that, as late as November 1879, the Corporation offered Rathmines a supply of a million gallons of Vartry water a day for an annual payment of £3,000. The Rathmines Commissioners turned the offer down, claiming that the township's Dodder supply would be better, cheaper, more plentiful and at higher pressure.

When the Rathmines Bill passed into law on receiving Royal Assent on 6th August 1880, Hassard and Tyrrell (whose offices were in Westminster) were appointed to oversee and bring the scheme to completion. The details were comprehensively set out in the Act as "described in the deposited books of reference." The Rathmines Commissioners were already well acquainted with Hassard and Tyrrell, who had also designed and supervised construction of the excellent Rathmines and Pembroke Main Drainage Scheme that came into operation in 1881. The equally successful reservoirs at Bohernabreena and the Dodder water supply are a further tribute to the ingenuity and competence of Victorian engineering.

RATHMINES WATER WORKS BOHERNABREENA

The Bohernabreena Reservoirs, in the valley of Glenasmole (The Valley of the Thrushes), lie at the foot of Kippure Mountain [MAP NO. 61 AND FIGS. 103, 104 & 105]. They are nine miles (15km) from the General Post Office in Dublin city centre and seven (11.5km) from Rathmines. There is a paradox here in that, as pointed out by Michael Murphy, no part of the reservoirs is located in the townland of Bohernabreena.

Unlike Dublin's Vartry scheme that predated it, Bohernabreena was designed not just to supply domestic water. Robert Mallet had presciently set out its other objectives more than thirty years earlier: "the combined purposes of providing an unfailing and increased supply of water-power to the mill

FIG. 103
Aerial photograph of
Bohernabreena Reservoirs
(Peter Barrow, 2014).

RATHMINES WATERWORKS. 10548. W.L.

RATHMINES WATERWORKS. 10547. W.L.

owners....and, contingently, of controlling the floods which have heretofore, at frequently recurring intervals, proved so destructive to property situated on its banks; and, as a further consequence giving the power of reclaiming those tracts of lands along the river banks, now of small or of no value, from periodical inundations; and lastly, preventing the annihilation of land in progress, by the degrading effects of these floods upon the clay banks of the river."

Hassard and Tyrrell were charged with providing a water supply to the township and a minimum discharge (compensation water) to the River Dodder to ensure a constant supply to the mills. Their first task was to identify the sources of clear water to be used for drinking and the coloured bog water that would become the compensation supply. As already noted, in the nineteenth century the technology did not exist to remove colour from bog water. Therefore, the principle of construction adopted at Bohernabreena was the method known as the separation principle.

The upper part of the Dodder catchment below Kippure Mountain and having an area of 4,340 acres (1,756 hectares) of granitic formation, was covered in peat; coloured brown in FIG. 106. Below this peat covered district there was an area of 3,250 acres (1,315 hectares), mostly uncultivated mountain and free from peat; coloured blue in FIG. 106. Partly of granitic but principally metamorphic and Silurian formation, it produced water of great purity; and owing to the occurrence of some gravel beds on the hill sides, the yield of spring water was especially bountiful. It was evident therefore that if the water from the peat-covered area could be intercepted and passed through the lower district without mingling, a supply of excellent water could be secured for township use. Achieving this happy objective was an integral part of the completed works at Bohernabreena.

THE RESERVOIRS

There are two impounding reservoirs at Bohernabreena (Glenasmole) — the Upper and the Lower [FIGS. 106, 107 & 108]. The Upper impounded the pure water chiefly for township (domestic) supply, while the lower one received the peaty water for mill owners' use only. It was originally intended to construct a reservoir of larger storage capacity, having its dam further down the valley from the present "lower" dam to impound the water from the peat covered district. This reservoir would have given a much increased supply for the mill-owners' use. It was proposed to tax them per foot of occupied fall, as had been done in other places. The mill owners objecting to this, an amendment was made during the progress of the Bill through parliament.

It was decided that the water from the upper district should be diverted past both reservoirs, except when the volume of the stream flowing from it exceeded 13.5 mgd (million gallons per day). This was calculated to be the maximum the mills could use and the Rathmines Commissioners were empowered to impound all water in excess of this flow. With this objective, works and gauges were provided to ensure that 13.5 mgd of water should pass down for the mill owners' use before any water from the upper district could flow into the reservoirs.

FIG. 104

Upper reservoir, Rathmines Waterworks, Bohernabreena (Robert French, circa 1900)

FIG. 105

View from embankment of upper reservoir to lower reservoir, Rathmines Waterworks, Bohernabreena, (Robert French, circa 1900)

187

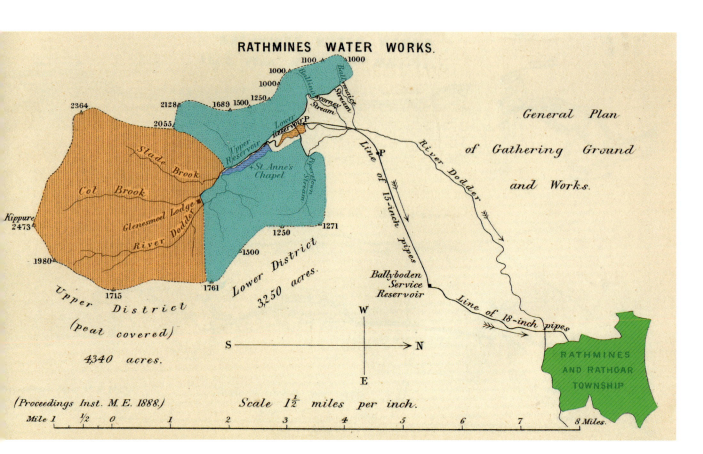

RATHMINES WATER WORKS.

General Plan

of Gathering Ground

and Works.

(Proceedings Inst. M.E. 1888.)

Scale 1½ miles per inch.

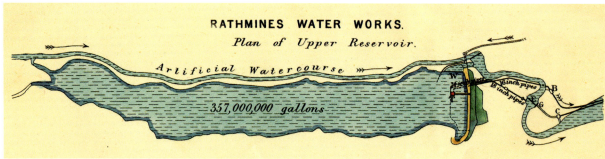

RATHMINES WATER WORKS.

Plan of Upper Reservoir.

Artificial Watercourse

357,000,000 gallons

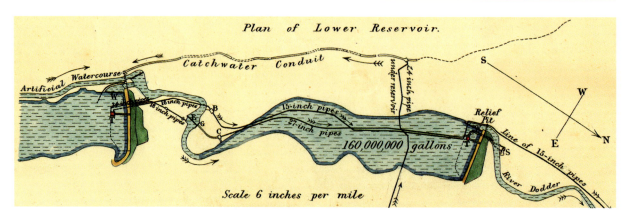

Plan of Lower Reservoir.

Catchwater Conduit

160,000,000 gallons

Scale 6 inches per mile

From weir "E" in FIG. 108 immediately above the lower reservoir the millers' preferential supply of 13.5 mgd was conveyed partly in open conduit "G" to "C" and partly through a line of iron pipes 27 inches in diameter "C" to "W". These were laid through the lower reservoir, terminating in the old course of the river below the reservoir embankment. Consequent on this arrangement the lower reservoir was constructed of less extent and capacity than originally intended.

Varying statistics as to catchments and capacities have been given over the years and those quoted here are from the 1986 River Dodder Flooding Report. The upper or clear water reservoir has a surface area of 57 acres (23 hectares), and a catchment area of 1,715 acres (695 hectares), with an embankment 656 feet (200 metres) long and 69 feet (21m) high. Designed to hold up to 360 million gallons (1,636 million litres), with its top water level 578 feet (174m) above Ordnance Datum, silting during a period of a hundred years reduced its capacity to 343 million gallons (1,559 million litres). The valve tower allows water to be taken at three different levels, 16 feet (4.8m), 34 feet (10.4m) and 52 feet (15.8m).

The lower reservoir has a catchment area of approximately 5,200 acres (2,104 hectares) and a surface area of 30 acres (12.1 hectares). The embankment is 590 feet (180 metres) long and 56 feet (17 metres) high. Its original capacity of 156 million gallons (709 million litres) was reported as reduced to 110 million gallons (500 million litres) in 1986, the top water level being 495 feet (151 metres) above datum. This reservoir had catchwaters, a gauge basin and duplicate measuring flumes for preferential and compensation water supplies to mill owners, whom the Rathmines Commissioners were required to service in times of drought with a minimum of 14.7 million gallons per week.

As the mills were closed on Sundays the minimum daily supply was 2.45 million gallons. This supply is maintained to the present day for fishing purposes and the method of monitoring is by measuring the flow over a rectangular weir at the Lower Reservoir in the basin known as the Millers' Gauge.

From this lower reservoir, compensation water could be released to maintain an adequate flow in the river during dry periods, allowing continuous operation of the mills. As the mills along the Dodder resorted to other forms of motive power, declined in number and eventually ceased to operate, the lower reservoir at Bohernabreena became a compensatory supply to the river. It is also a flood relief buffer as originally conceived by Robert Mallet and brought into existence by Richard Hassard. There are recorders in both reservoirs and in the by-pass canal, from which accurate information is compiled about reservoir levels and flows over the spillways.

The very heavy flash floods that periodically affect the valley have caused several landslips, and a characteristic of the catchment is that the water flows off the slopes very quickly. The overflow weirs were therefore made 200 feet (60m) in length to deal with the excess water.

FIG. 106

General plan of gathering ground and works" showing the catchment areas for drinking water (blue) and millowners' compensation water (brown) (Authur Tyrrell, 1888)

FIG. 107

Plan of upper reservoir (Tyrrell, 1888)

FIG. 108

Plan of lower reservoir (Tyrrel, 1888)

THE EMBANKMENTS

When forming an earthen gravity dam the first task is to clear the floor and the slopes of the valley where the reservoir is to be sited. In the vicinity of the embankment the area is excavated down to the boulder clay. Next the eduction tunnel is formed in the boulder clay as detailed in **FIG. 109**. The river is then diverted into the eduction tunnel so that the construction of the embankment can proceed. Across the valley floor a trench is dug and filled with puddle clay. The puddle clay core prevents water flowing through, under and at the sides of the dam. In most earthen gravity dams the puddle clay core is taken down to bedrock. The Glenasmole Valley is unusual in that there was a large depth of boulder clay overlaying the bedrock. In the dams in the Glenasmole Valley the puddle clay core was constructed to a sufficient depth so as to prevent water seeping under the dam.

The next sequence is normally the construction of the embankment in approximately 2 metre lifts. In Glenasmole, on the eastern slopes, there was a layer of sand and gravel, about 37 metres wide, covering the boulder clay. A concrete cut off barrier had to be formed as detailed in **FIG. 110** to prevent water seeping around the side of the embankment. If water seeped around the earthen gravity dam the sand and gravel would eventually be washed away leading to a catastrophic failure of the dam. This concrete cut off barrier had to be completed before the embankment construction could proceed. **FIG. 111** shows a typical earthen embankment, at base level, under construction with a puddle clay core. **FIG. 112** shows typical earthen embankment nearing completion. **FIG. 113** shows a typical reservoir construction site in the mid-nineteenth century. Note the material being brought up the slope in a wheelbarrow.

The cross-section in **FIG. 114** shows the central puddle core with the eduction tunnel running through the embankment.

Arthur Tyrrell described the construction of the Bohernabreena earthen gravity dams with puddle clay to stop water seeping through the embankments. "The two embankments were formed with slopes of 3 to 1 on the inner and 2 to 1 on the other faces [**FIGS. 114, 115 & 116**]. On both however there occurred on the eastern side of the valley veins and deposits of sand and gravel extending for a considerable distance into the hill, and rendering it necessary to follow them in by headings driven one over another and filled with concrete, the ground being too steep for open trenches. This operation was one of considerable difficulty, owing to the loose nature of the sand and the quantity of spring water encountered. The lower heading at the lower embankment was extended 120 feet into the hill-side, until the deposit of sand had been completely gone through and a wall of hard blue clay reached.

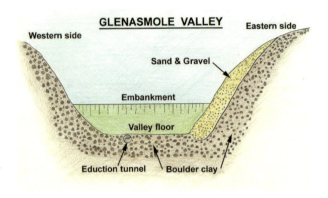

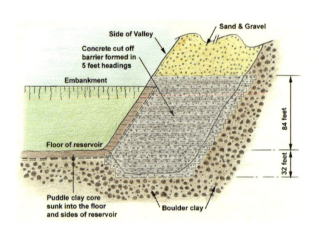

FIG. 109

Section through
Glenasmole Valley at reservoir
embankment (Aisling Walsh
and Don McEntee)

FIG. 110

Detail of concrete cut off
barrier at side of embankment
(Aisling Walsh and
Don McEntee)

FIG. 111

Typical gravity dam with puddle clay core during construction

FIG. 112

Typical reservoir embankment under construction

FIG. 113

Typical reservoir under construction

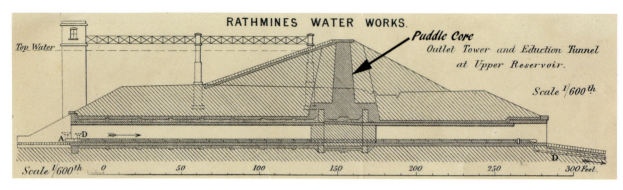

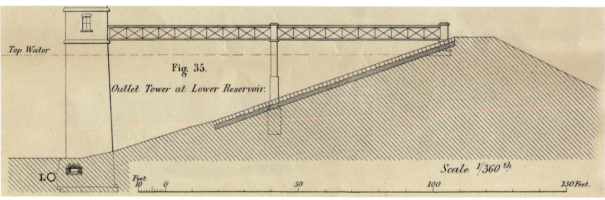

FIG. 114
Cross section through dam of
upper reservoir (Tyrrell, 1888)

FIG. 115
Upper slope of dam and outlet
tower of lower reservoir
(Tyrrell, 1888)

FIG. 116
Embankment to lower
reservoir (Don McEntee)

FIG. 117

Heading into side of a
hill to form concrete
cut off barrier

FIG. 118

Hermaphrodite cart used
on construction sites in
nineteenth century

"Into this clay the heading was carried sufficiently far to ensure a sound junction being effected between the concrete filling and the impervious material at its termination. From the floor of the lower heading a trench was then sunk to a depth of 32 feet, when the hard clay was reached; and the whole was refilled with carefully rammed concrete. After this the next heading was proceeded with, and filled with concrete in the same manner; and so on until a height of 84 feet from the bottom was attained, when the work was completed by a short length of open trench.

"The headings being only 5 feet in height and all heavily timbered, this part of the work was necessarily very slow and tedious, greatly retarding the progress of the upper reservoir embankment and consequently the final completion of the works."

FIG. 117 shows a heading being driven into the side of a valley to form a cut-off barrier at the side of the reservoir and the conditions under which the construction workers had to put up with. FIG. 118 is an example of a cart used on many construction sites in the nineteenth century. It was known as a hermaphrodite cart. The cart had three wheels for manual manoeuvring by one or two men. The front wheel could be removed and a horse (or donkey) harnessed to permit use as ordinary horse (or donkey) and cart transport.

EDUCTION TUNNELS AND VALVE TOWERS

Andrew Tyrrell wrote: "The eduction tunnels, through which the River Dodder was diverted during the progress of the embankments, are constructed in the solid ground on the western side of the valley and entirely below the puddle trenches. They are in each case 11 feet in diameter, having a sectional of 100 square feet and with the exception of the central plugging, which is of brickwork, are built entirely of rubble masonry backed with concrete.

The outlet towers to the reservoirs are in close proximity to the forebays of the tunnels and reached by a footbridge from the embankments [FIGS. 114 & 115]. There are three openings, marked "N" in FIG. 119, at the outlet tower of the upper reservoir through which the water is drawn off at different levels for the township supply. From the base of the tower, extending through the eduction tunnel to the valve chamber, are laid two lines of pipes as shown in FIG. 120 (upstream of the plug in the eduction tunnel), one of 24 inches diameter for emptying purposes and the other 16 inches diameter for the water supply to the township. A diagrammic layout of the pipes from the Upper Resevoir is shown in FIG. 121.

An 18-inch pipe was installed (through the plug in the eduction tunnel) to provide drainage through the upper reservoir dam when the eduction tunnel [FIGS. 122 & 123] was being sealed with the plug. This pipe was also be used to draw down the water levels in the upper reservoir when a storm was forecast. Now the 18-inch pipe is permanently sealed. The 24-inch and 16-inch pipes are controlled, in addition to the ordinary valves at the exit of the eduction tunnel, by stop plug valves [FIGS. 124, 125 & 126] suspended from a crab winch "H" [FIG. 119], and guided on to their seats by a specially designed arrangement in the form of a large bell-mouth attached to the upturned brass plate [FIG. 124].

The piping layout prior to 2006 is shown in FIG. 121. The original 16 inch pipe from the Bohernabreena valve tower has sufficient head to deliver four million gallons per day. Beyond the eduction tunnel the discharge of the 16-inch was split between the 12 and 16 inch pipes (now a single 600mm pipe)

As shown in FIG. 121 water could be drawn from the valve tower in a 24-inch pipe discharging to the spillway. This served two purposes. If the lower reservoir ran dry, water could be extracted from the upper reservoir to the lower reservoir to ensure a constant supply to the millers. When a large storm was forecast the standard operating procedure was to draw down the water levels in both reservoirs to provide storage for the flood waters. This prevented or delayed flooding in the lower reaches of the River Dodder.

The three pipes at the exit of the eduction tunnel to the upper reservoir are shown in FIG. 122.

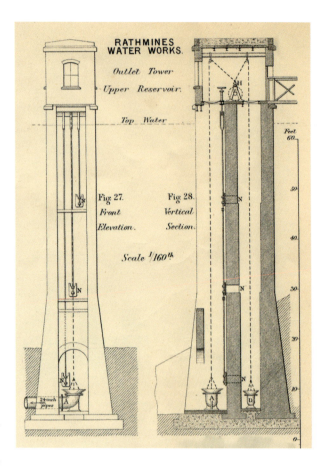

FIG. 119

Outlet tower, upper reservoir with plug valves to outlet pipes at "A" & "D" (Tyrrell, 1888)

FIG. 120

Section at entry to eduction tunnel, upper reservoir (Tyrrell, 1888)

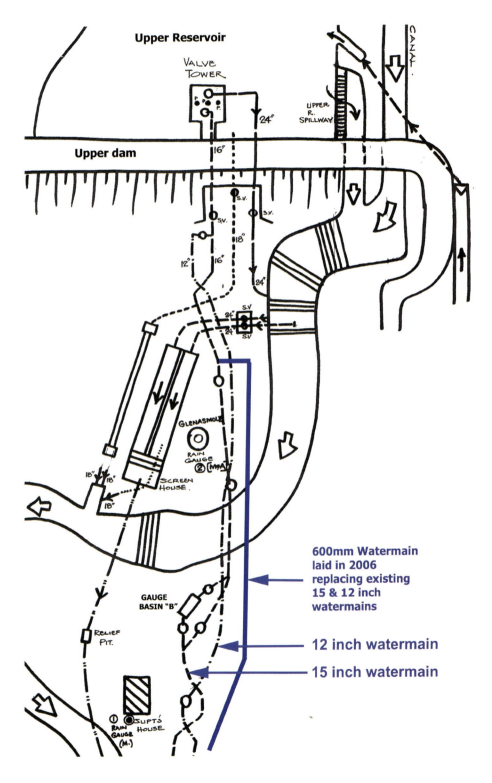

Upper Reservoir

VALVE TOWER

24"

UPPER R. SPILLWAY

CANAL.

Upper dam

16"

S.V.

S.V.

S.V.

18"

12" 16"

24"

24" S.V

24" S.V

GLENASMOLE

RAIN GAUGE ② (MMS)

18" 18"

18"

SCREEN HOUSE.

600mm Watermain laid in 2006 replacing existing 15 & 12 inch watermains

12 inch watermain

GAUGE BASIN "B"

RELIEF PIT.

15 inch watermain

RAIN GAUGE (M.)

SUPTS HOUSE

FIG. 121

Schematic layout of pipe work at upper dam and reservoir, Bohernabreena (1986)

FIG. 122 (LEFT)
16, 18 & 24 inch pipes in
eduction tunnel outlet, upper
reservoir (Don McEntee)

FIG. 123 (BELOW)
Entrance to eduction tunnel,
upper reservoir (Don McEntee)

FIG. 124 (TOP LEFT)
Section through plug valve
(Tyrrell, 1888)

FIG. 125 (BOTTOM LEFT)
Plug valve stored in takeoff
tower (Don McEntee)

FIG. 126 (BOTTOM RIGHT)
Plug valve being lowered
onto an outlet pipe at side of
takeoff tower

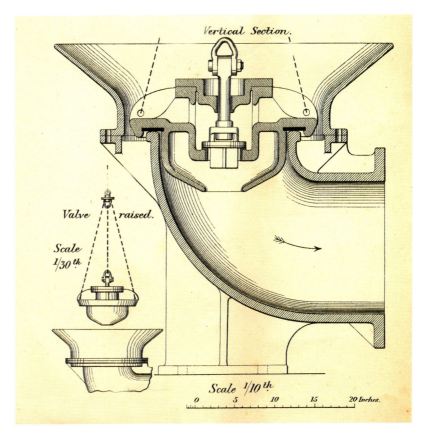

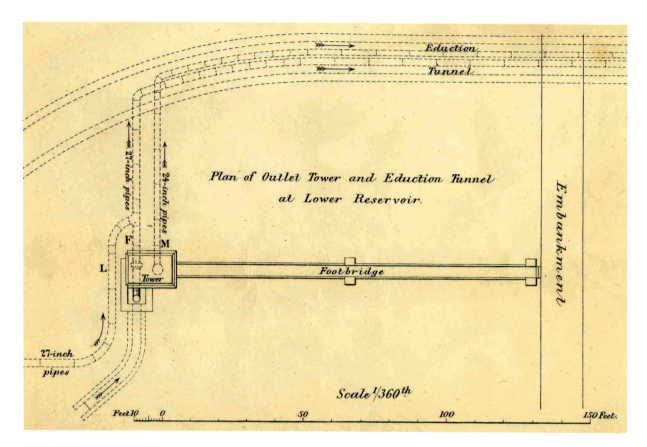

Plan of Outlet Tower and Eduction Tunnel at Lower Reservoir.

1600 mm OUTLET

"From the outlet tower of the lower reservoir are laid two lines of pipes, of 27 and 24 inches in diameter [FIG. 127]; the larger forms a junction with the line of the 27-inch pipes laid through the reservoir, which conveys the water from the upper district; and the smaller for emptying purposes, also extends through the eduction tunnel and to a valve chamber and continues to the mill-owners' gauge basin in a 12 inch pipe". From the valve chamber the 24 inch pipe discharged into a pit before spilling to the lower spillway. The outlet from the 27-inch pipe was extended during the reconstruction of the lower spillway to the pit in which the 24-inch pipe also discharges.

A major deficiency with the original pipe layout from the lower reservoir was the connection of the 27-inch preferential flow pipe from the upper district into the 27-inch pipe coming from the outlet tower. For the purposes of flood control the operating procedure was to draw down both reservoirs when a storm was approaching the valley. At this stage the 27-inch preferential supply pipe from the upper district would be flowing full and discharging directly into the river below the lower reservoir. The draw-down water from the upper reservoir flows into the lower reservoir. As the pressure head on the 27-inch pipe laid in the bed of the lower reservoir is greater than the head of water in the lower reservoir no water would be drawn down from this reservoir in the 27-inch pipe from the outlet tower.

In the conditions described only the 24-inch pipe functioned as a means of drawing down the lower reservoir. This deficiency was partly rectified in the 2006 upgrade of the waterworks with the provision of a 1600mm diameter valve outlet directly into the reconstructed lower spillway [FIG. 128]. Then the top four metres of the lower reservoir can now be quickly drawn down in storm conditions.

STREAMS, WATERCOURSES, CHANNELS, DIVERSIONS

The network of streams, watercourses, channels and various diversions around Bohernabreena is extensive and complicated [MAP NOS. 2, 3 & 61].

The Dodder upstream of Castlekelly drains the peat covered areas below Kippure. A new artificial watercourse [MAP NO. 3], intercepting the upper Dodder was constructed from above Old Castlekelly Bridge to the bywash channel at New Castlekelly Bridge. This new watercourse intercepted the Cot Brook and the Slade Brook. At the New Castlekelly Bridge a sluice gate was installed so that in times of very low flows water could be diverted from the artificial channel into the Upper Reservoir. Also at the New Castlekelly Bridge a weir was constructed so that flood discharges in excess of the capacity of the bywash channel would be diverted into the Upper Reservoir.

To capture the clean water from Castlekelly a pipe was laid from the former bed of the Dodder into the head of the Upper Reservoir.

BYWASH CHANNEL AND UPPER SPILLWAY

The artificial watercourse [FIGS 107 & 129] that conveys the water from the upper River Dodder, the Cot Brook and the Slade Brook to the Lower Reservoir is known as the bywash channel and locally as the "Canal". It has a carrying

FIG. 127
Pipe layout at takeoff tower,
lower reservoir (Tyrrell, 1888)

FIG. 128
Weir and 1600mm outlet into
reconstructed stilling basin,
lower reservoir (Don McEntee)

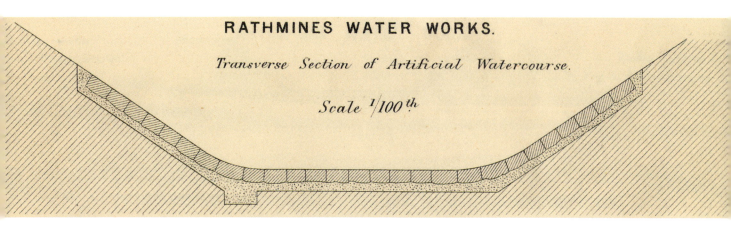

RATHMINES WATER WORKS.

Transverse Section of Artificial Watercourse.

Scale ¹/100ᵗʰ

FIG. 129
Cross section of bywash
channel (Tyrrell, 1888)

FIG. 130
Gauging Station on bywash
channel (Don McEntee)

FIG. 131
Weir at New Castlekelly
bridge directing river flow
from mountains into bywash
channel on left with overflow
into upper reservoir
(Don McEntee)

FIG. 132
Flood waters at New
Castlekelly bridge
overflowing the weir
(Ger Goodwin, 2008)

capacity of 120,000 cubic feet per minute (1080mgd or 56 cumecs) equal to a rate of about 11 inches of rain in 24 hours. This runs along the western bank of the upper reservoir [FIG. 107] and at the time of construction was cut into the side of the valley. The gauging station on the bywash channel is shown in FIG. 130. The water flowing from the upper or peat covered area of 4,340 acres, which was entirely unsuitable for township supply, was diverted past the upper reservoir in the bywash channel and discharged into the lower reservoir below the upper embankment.

The bywash channel had to deal with floods of considerable magnitude. In the event of any obstruction to the channel (landslips or other unforeseen circumstances) or flows in excess of the capacity of the bywash channel, the works adjoining the upper and lower embankments—overflow weirs, tumbling bays, bywashes and so on—were of sufficient capacity to carry off 30 cubic feet per acre per minute, or a rainfall of 12 inches in 24 hours. The capacity of the New Castlekelly Bridge was equal to that of the bywash channel with any excess water flowing over the weir into the box culverts and into the upper reservoir [FIG. 131]. FIG. 132 shows the flood waters cascading over the weir upstream of the bridge into the upper reservoir.

LOWER VALLEY DRINKING WATER CATCHMENT

To maximise the quantity of clean water available springs and streams, on the western side of the valley and adjacent to the Upper Reservoir, were conducted into it by culverts passing under the bywash channel. Lower down the valley, water from the Ballinascorney and Ballymaice streams, distant respectively 1¼ miles (2 km) and 2 miles (3.2km), were intercepted, the direction of their flows reversed and the water conveyed back into the Upper Reservoir by catchwater conduits (open watercourses interspersed with stone drains) laid along the hillsides [FIG. 133]. The stone drains were at some stage replaced with cast iron pipes.

On the eastern side of the valley the Piperstown Stream flowed down into the Dodder at Fort Bridge. This is the bridge that crosses the river Dodder near the entrance to the Waterworks on the Ballinascorney Road. Upstream on the Piperstown Stream, a dam was constructed to divert some of the water in the stream into a channel. The water was taken into this open channel and carried across from Glassamucky to a tower and siphoned under the lower reservoir in a 24-inch diameter cast iron pipe [FIG. 108]. The pipe discharged into the open channel from Ballinascorney and Ballmaice and into the upper reservoir. In time the siphon probably became blocked with material from the Piperstown River. There is no record of a flushing system installed and no method of scouring the pipe under the reservoir.

At some time before 1935 the 24-inch pipe probably ceased to function due to a blockage under the reservoir. As the clean water from the Piperstown Stream was not flowing into the upper reservoir a new 9-inch (225mm) pipe was laid across the top of the lower dam and connected into the 12-inch (300mm) watermain in the avenue. This connection was probably made when the 12-inch pipe was being laid in 1935 as described later. This pipe laid on top of the dam put the dam at risk. If the pipe burst on top of the dam it could have weakened

FIG. 133
Ballinascorney catchwater
channel to upper reservoir

FIG. 134
Schematic layout of pipes
at lower dam and reservoir,
(Bohernabreena, 1986)

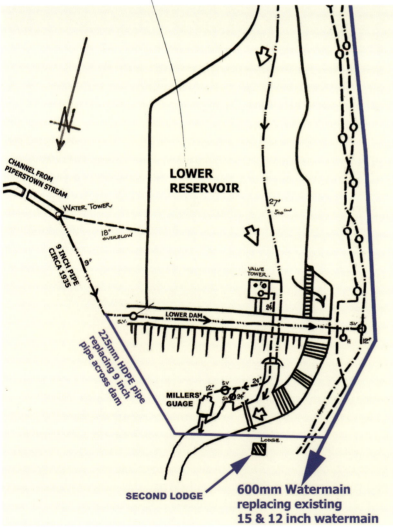

FIG. 135

Laying cast iron pipe in
Dublin Street using a gin

the structure and led to a catastrophic failure of the dam. When the works were
upgraded in 2006, a new 225mm pipe was laid down the eastern bank of the
valley, crossing under the river and connected to the new 600mm watermain.
The schematic layout of the pipework is shown in **FIG. 134**.

During the Second World War large diameter cast iron pipes were in short
supply. Just after the war the 24-inch diameter iron pipe was removed from both
sides of the valley for use in Benburb Street, Dublin. After the excavation of the
trenches the individual pipes, each weighing over one ton, were lifted by means
of a hand operated pulley system mounted on a tripod of wooden poles known as
a "gin" [**FIG. 135**]. To transport the pipes across the reservoir each pipe was lifted
on to a trio of clinker built boats across which a platform of wooden planks was
laid. Approximately 520 yards of 24-inch pipes were salvaged. At the same time a
deep trench was dug at the upper end of the reservoir and sections of the original
15-inch pipe, which had been laid under the reservoir, were removed for re-use.

BYWASH CHANNEL

The peaty water from the granite areas was channelled around the upper reservoir in the bywash channel and routed through the lower one in a 27-inch (685mm) diameter pipe. It was possible to divert some of the river water into the upper reservoir, if this was necessary, to maintain the water supply to Rathmines.

Because preserving the quality of the drinking water was of paramount importance, contaminated water was collected in a pipeline laid from Castlekelly through the valley to connect with another line from Cunard and siphoned into the bypass channel upstream of the gate lodge at Castlekelly.

During construction, part of the valley was unstable and a series of arches was constructed to stop the bank from sliding into the bywash channel [FIG. 136]. When there is a big flood in the river FIG. 137 large quantities of debris are carried down and lodge in the arches, blocking the bywash channel [FIG. 138]. The river then overflows its banks into the upper reservoir, causing damage to the avenue and the banks of the reservoir. The 200-foot (60m) long overflow weirs at the upper [FIG. 138] and the lower reservoir, [FIG. 140] already described, were designed to deal with exceptional flooding.

A short distance south of the arches three bell mouthed pipes (18 inches in diameter) were laid in the bed of the bywash channel and connected by three channels under the embankment into the upper reservoir [FIG. 141]. This was to divert flows during maintenance work in the bywash channel and divert the water from the bywash channel into the upper reservoir, if necessary, to maintain the water supply to Rathmines. A sluice gate was later installed in the overflow weir at New Castlekelly Bridge so that water from the river could be diverted into the upper reservoir.The inlets to these channels are through portholes in the bed of the river and under normal operating conditions are plugged. During the construction of the new upper spillway in 2004 these channels were used to divert the flow from the bywash channel into the upper reservoir.

FIG. 136
Arches on bywash channel to prevent ground on left slipping into upper reservoir (Don McEntee)

FIG. 137
Flood flow through arches
on bywash channel
(Ger Goodwin, 2008)

FIG. 138
Arches blocked
with debris

FIG. 140
Original weir to lower
reservoir (Don McEntee)

FIG. 139
Original weir to upper
reservoir (Don McEntee)

FIG. 141
Port holes in bywash
channel (Ger Goodwin)

FIG. 142

Pitched stone in bywash channel (Don McEntee)

During the construction of the upper spillway a man was walking with his dog on the avenue beside the bywash channel. While the dog was running in and out of the bywash channel, the man suddenly became aware that the animal had stopped running back and forth. He began calling and searching, fearful that the dog had run off after sheep in the adjoining lands. A short time later he heard barking from the reservoir and found his dog in a distressed state. While running in the bywash channel the dog was sucked into one of the pipes in the bed of the channel and washed into the reservoir. But, on being reunited with his master the dog was none the worse for his adventure.

A typical view of the bywash channel is shown in FIG. 142. The bed and side slopes are constructed using random rubble stone. At the end of the bywash channel the flow in the channel takes a right-hand turn, joins the flow from the upper reservoir and flows down the upper spillway. To prevent scouring, the designers came up with an ingenious method of turning the flow in the river. [FIGS. 143, 144 & 145]. The objective was, in times of heavy floods, to prevent the great body of water impinging on the side walls of the tumbling bay in consequence of the direction of the current being suddenly turned at right angles. Also checked would be the rush and swirl of water which otherwise at such times would occur in the diverted channel at the point of its junction with the tumbling bays and bywash.

This precautionary work fulfilled its purpose admirably. The twenty culverts through the weir were capable of discharging not only the ordinary flow of the river but considerable floods. At times of excessive rainfall, when the bywash

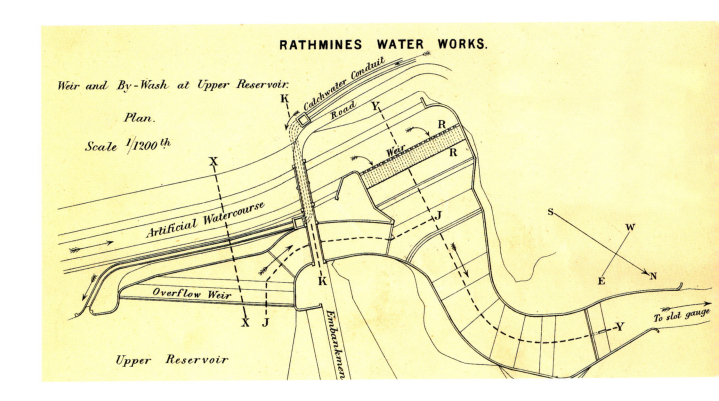

FIG. 143
Slotted weir in artificial
watercourse (bywash channel
at upper reservoir. Upper
spillway shown from J toY
(Tyrrell, 1888)

FIG. 144
Section RR through slotted weir
(Tyrrell, 1888)

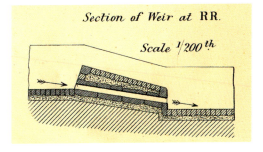

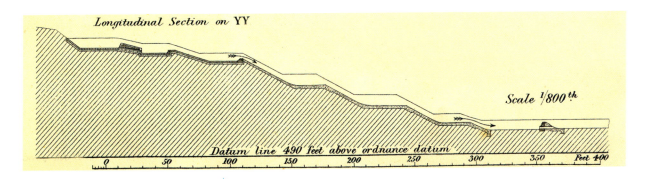

FIG. 145
Longitudinal section YY
through spillway (Tyrrell, 1888)

FIG. 146
Slotted weir and bridge at bywash channel (Don McEntee)

FIG. 147
Footbridge & outlet from 3 inch pipe (Michael Murphy)

channel was running nearly full, the surplus water passed quietly away over the crest of the weir, leaving comparatively still water behind it, the velocity of the water in the channel being no greater at this point than in any other part of the artificial watercourse. Details of the weir are shown in FIG. 146.

To make sure that the water in the upper reservoir would not be contaminated, the designers built a collection system at the end of St. Anne's Graveyard on the eastern bank of the reservoir. Michael Murphy notes that a tender was issued for the laying of 627 lengths of 3-inch cast iron pipes in July 1897. These pipes took the discharge from St. Anne's Graveyard along the eastern side of the reservoir to a manhole located at the lower eastern corner of the Upper Dam and discharged into the Dodder at the southern end of the footbridge located behind the Superintendent's house An iron deposit can be seen in the river at this location [FIG. 147].

FIG. 148
Waterworks boundary marker
(Don McEntee)

WATER SUPPLY FOR THE RATHMINES TOWNSHIP

The Upper Reservoir [MAP NO. 61] impounds water chiefly for township supply. In the event of the lower reservoir running dry the millers' compensation water would be drawn from this upper reservoir. The basis of sizing the two reservoirs was to ensure that the pure water would not be used for millers' compensation water.

An interesting feature of the Bohernabreena complex, illustrated in Christopher Moriarty's *Down the Dodder*, was the erection by the Rathmines Commissioners of boundary stones at the boundary of the clean water collection area shown in blue (Lower District) in FIG. 106 . The Rathmines boundary stones displayed a single letter W [FIG. 148]. These stones were similar to those used by the War Office to delineate the extent of military property.

As explained by Tyrrell, a supply of excellent water could have been secured for the township from the lower part of the valley. In 1880 the water as taken from the streams in dry weather contained about four grains of solid matter per gallon, and was about five degrees of hardness: "a more desirable water would have been difficult to procure."

The streams and springs on the western side of the valley, adjacent to the upper reservoir, were conducted into the reservoir by culverts passing underneath the artificial watercourse; the construction of these culverts is shown in FIG. 149. In the lower part of the valley, the waters of the Ballinascorney and Ballymaice streams [FIG. 106], distant respectively 1.5 and 2 miles from the Upper Reservoir, were intercepted. The direction of their currents was reversed and the water was conducted back into the upper reservoir by catchwater conduits constructed along the hillside.

As shown in FIG. 150, the catchwater conduit was lined with random stones. Before this catchwater conduit [FIG. 151] reached the bywash channel it flowed into a triple stone culvert passing under the bywash channel and discharging into the upper reservoir south of the weir. At the entry into the culvert [FIG. 152] a vee notch weir measured the flow into the reservoir. Upstream of the vee notch weir the channel was sized to the capacity of the culvert and any excess water was spilled over a side weir for discharge into the spillway and the lower reservoir [FIG. 153]. Details of the triple culvert under the roadway are shown in FIGS. 154, 155 & 156.

FIG. 149
Culvert under bywash channel
(Tyrrell, 1888)

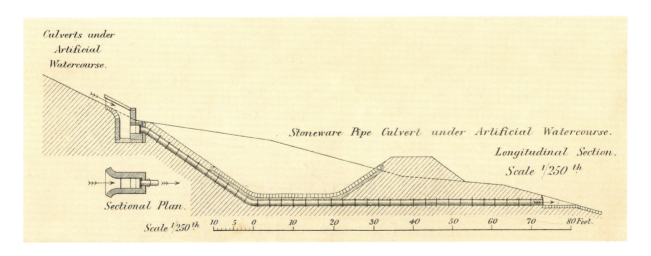

FIG. 150
Ballinascorney
catchwater channel
(Don McEntee)

FIG. 151
Catchwater channel
entering culvert
(Don McEntee)

FIG. 152
Weir to measure flow at entry
to treble box culvert
(Don McEntee)

FIG. 153
Side weir overflow to
catchwater channel upstream
of treble box culvert
(Don McEntee)

Tyrrell recorded that all the masonry in the storm channels, overflow weirs, eduction tunnels, valve towers, bridges, sheeting of the embankments, and works generally, was common rubble stone found in the locality. It was mixed with Portland cement mortar in the proportion of one part of cement to three parts sharp sand, and the concrete, of which a large quantity was used, was composed of excellent gravel. This was brought down by the floods, found in the valley, and mixed in proportion of one part of cement to six parts of gravel. No lime was used in any part of the works.

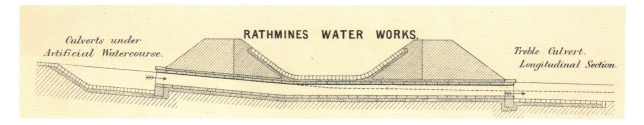

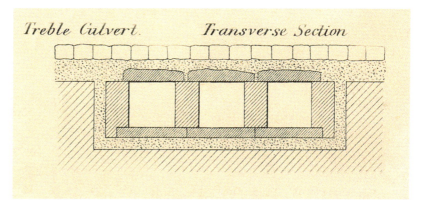

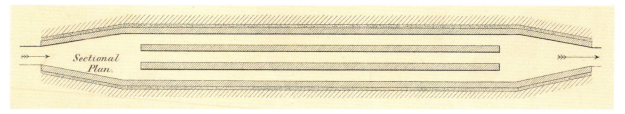

FIG. 154

Treble culvert under bywash channel (Tyrrell, 1888)

FIG. 155

Section through treble culvert (Tyrrell, 1888)

FIG. 156

Sectional plan of treble culvert (Tyrrell, 1888)

MILLERS' COMPENSATION AND PREFERENTIAL SUPPLIES

The Rathmines and Rathgar Water Act contained several sections specifically worded to protect the rights of mill owners and other existing water users. Despite such reassurances, the mill owners, understandably exasperated by years of uncertainty over water supplies, were not convinced that the Bohernabreena scheme would improve their lot. Their vociferously expressed fears that the works as carried out would seriously worsen their interests worried the Rathmines Commissioners, who could be sued if the water supply turned out to be inadequate or reduced. They therefore consulted Sir John Hawkshaw, the eminent engineer who was very familiar with Dublin water matters, and who found the situation would be much improved, with satisfactory provision for the millers.

The Rathmines Water Act stipulated that when the flow in the river was below normal "the Commissioners shall cause to flow into the River Dodder for the use of the mill owners in the course of every week a quantity of water not less than 14,700,000 gallons at a rate of 2,450,000 gallons on each of the six working days of 24 hours would pass in a regular and continuous flow from the Miller' gauge into the Dodder. This compensation water was taken down through the take-off tower [FIG. 157] of the Lower Reservoir to a gauging basin [FIG. 158]. It should be noted that the Millers' Gauge was constructed on the right bank of the river below the lower dam [FIG. 134] and not as shown (position S) on the original drawing [FIG. 158].

When Michael Murphy was living in the Waterworks one of this father's duties, as Caretaker, was to turn on and off the supply of water to the millers' gauge. It was cut off every Saturday at about mid-day and turned on again on Sunday at about 6.00 pm. Now the millers' gauge is left running continuously and the graduated brass scale that was beside the outlet has long since disappeared.

During times of normal flow the Act required that the preferential supply of 13.5 million gallons per day was diverted past the lower reservoir. A weir "E" [FIGS. 108 & 159] was constructed to divert the preferential supply into a 27-inch pipe laid in the bed of the lower reservoir [FIG. 108], discharging within the eduction tunnel into the River Dodder below the lower reservoir [on left of photo in FIG. 161]. A plan of the weir is shown in FIG. 159 and a photograph in FIG. 160 indicating the location of the slot weir which led to the start of the 27 inch pipe at "C" in FIG. 108.

The preferential supply passed through the slot gauge "G" in FIG. 108, which was capable of delivering the stipulated quantity with a mean head of 9 inches. This was the height of the crest of the adjoining weir [FIG. 159]. An elevation and section of the slot gauge to a large scale are shown in FIGS. 162, 163 & 164. From the slot-gauge the water flowed down an open conduit about 100 yards long into the beginning of pipes 27 inches in diameter, by which it was conveyed under the lower reservoir into the old river course, according to the Act of Parliament. At the head of the 27-inch pipe ["C" IN FIG. 108] was placed a sluice with a weir adjoining [FIGS. 165, 166 & 167]; the object of placing the sluice in this position was to regulate the flow so that only the stipulated quantity of 1,500 cubic feet per minute passing through the slot-gauge under a head of 9 inches should be admitted into the 27-inch pipe The surplus in times of flood, when with an increased head the gauge passes down more than the required quantity of water, was discharged over the weir at the head of the 27-inch pipe, and into the lower reservoir.

It can be seen that with this arrangement of the slot gauge the engineers went to great lengths to ensure that only the quantity of preferential water set out in the Act would be permitted to flow down the river in times of normal flow. All extra water would be stored in the lower reservoir for mill-owners' compensation during periods of low flow. At times of high discharges from the Upper Dodder the lower reservoir filled up to the level of the weir and then discharged freely over it, down the lower spillway and back into the old river channel.

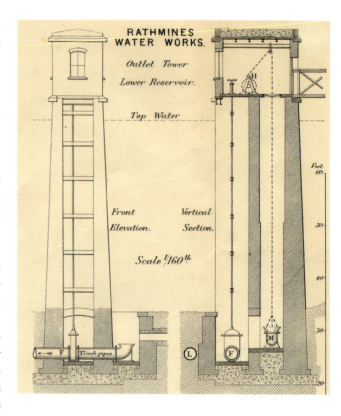

FIG. 157
Outlet tower to lower reservoir
(Tyrrell, 1888)

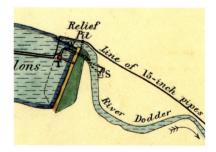

FIG. 158
Original design location of gauging basin at 'S' on left bank of Dodder. (The constructed position of miller's gauge is on right bank - Fig. 134) (Tyrrell, 1888)

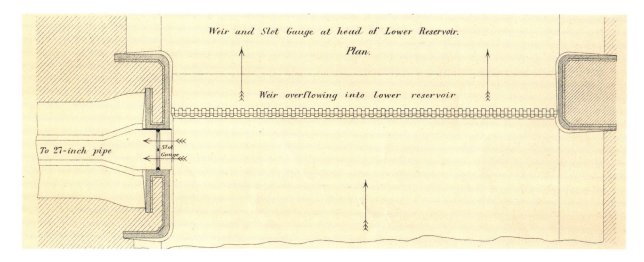

FIG. 159

Weir and slot gauge at head of lower reservoir – weir 'E' and slot guage 'G' in Fig. 108 (Tyrrell, 1888)

FIG. 160

Weir 'E' below upper spillway (Don McEntee)

FIG. 161

Exit from lower reservoir eduction tunnel on left and lower spillway on right before spillway upgrade (Don McEntee)

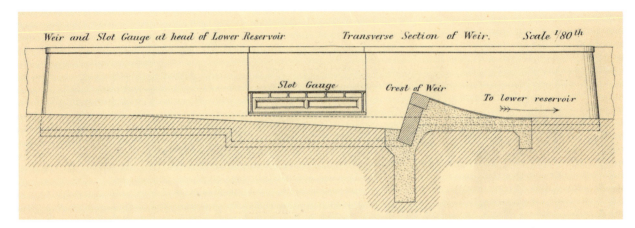

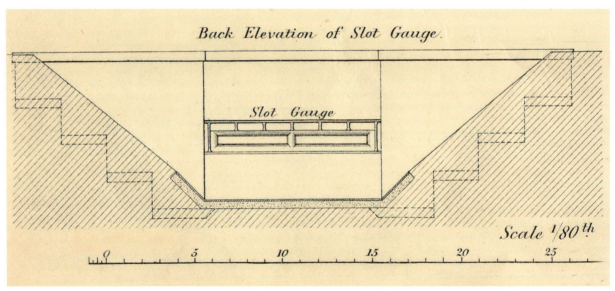

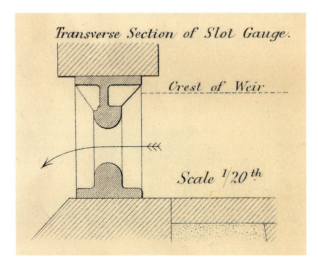

FIG. 162
Weir and slot gauge leading to
original entry to 27 inch pipe under
lower reservoir (Tyrrell, 1888)

FIG. 163
Back elevation of slot gauge
(Tyrrell, 1888)

FIG. 164
Transverse section of slot gauge
(Tyrrell, 1888)

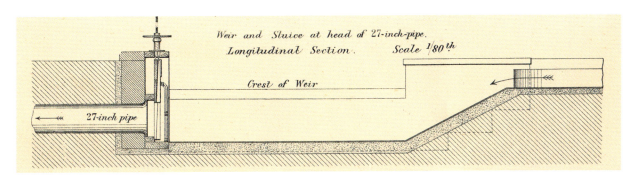

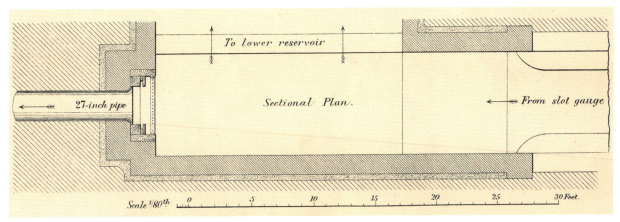

FIG. 165
Weir and sluice at head of
27 inch pipe passing under lower
reservoir (Tyrrell, 1888)

FIG. 166
Sectional plan of weir at head
of 27 inch pipe (Tyrrell, 1888)

FIG. 167
Transverse section of
Fig. 166 (Tyrrell, 1888)

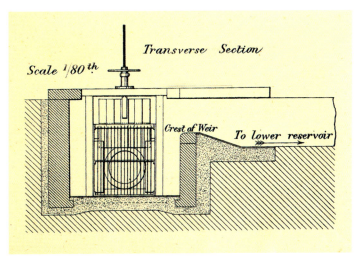

FIG. 168 (LEFT)
Miller's gauging basin
(Don McEntee)

FIG. 169 (BELOW)
Weir at outlet to millers'
gauging basin
(Don McEntee)

FIG. 170

Revised system for
preferential flow to 27 inch
pipe under lower reservoir.
Water abstracted from lower
part of spillway discharged
to twin channel.

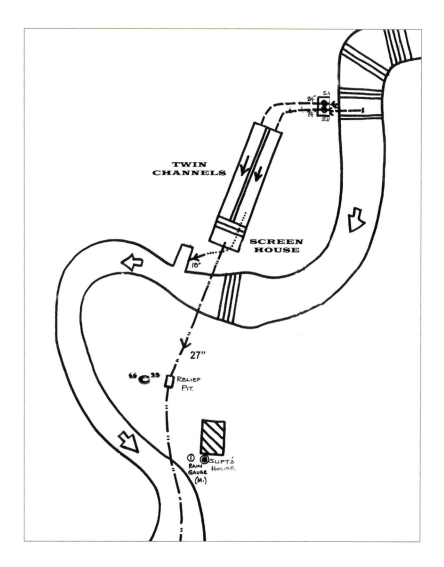

In times of drought when there was no flow coming down the Dodder
from the upper catchment, water was drawn from the lower reservoir in a
24-inch pipe with a 12-inch connection to the millers' gauging basin below
the reservoir [FIG. 134]. At the lower draw-off tower water could be taken at
three levels depending on the level in the reservoir. This water was piped to
the millers' gauging basin [FIG. 168]. The flow was adjusted so that the level in
the basin was 9 inches over the outlet weir [FIG. 169]. This kept the flow at the
rate stipulated for mill-owners' compensation at 2.45 million gallons per day.

After the waterworks opened in 1887 a problem occurred with the system
for the preferential supply. It is probable that the slot-gauge was being clogged
continually with debris carried down the river. A new system for improving the
preferential supply was therefore constructed. The date of this alteration to the
original design is not recorded, but the solution was every bit as ingenious as
the original design.

FIG. 171

Upper spillway, Rathmines
Waterworks, Bohernabreena
(Robert French, circa 1900)

A new take-off point was constructed near the bottom of the upper spillway [FIGS. 170 & 171]. The flow from the spillway passed through a double valve [FIG. 172] and discharged into a twin open channel [FIG. 173]. A cog system joined the original valves so that as one closed the other one opened. This was to ensure that the flow was in the region of 13.5 million gallons of water per day. Flows in excess of this would have caused local flooding downstream at the screen house. In the channel large debris that passed through the valve settled out as shown in the right channel in FIG. 173. When one channel was in use the other channel could be cleared of accumulated debris. At the end of the twin channel the flow passed over a weir and entered the screen house [FIG. 175]. Nine inches of water over the weir approximated to the stipulated 13.5 mgd.

In the screen house the smaller debris was removed before the water discharged back into the original 27-inch pipe "C" [FIGS. 108 & 170]. Upstream of the screen house, weirs were constructed at the sides of the channel [FIG. 174]. Flows in the channel in excess of 13.5 mgd passed over the weir and into an 18-inch diameter pipe discharging into the river channel. A control weir was constructed at the outlet of the 27-inch pipe at the exit of the eduction tunnel below the lower reservoir [FIG. 161] to check that only the preferential supply was discharging to the river.

The design of the screen [FIG. 175] was simple and effective. The water flowed over a weir into the house, inside which two screen walls were built with a series of 2-inch holes in layers [FIG. 176]. When the first screen wall became blocked with debris, the water overflowed into a transverse channel

FIG. 172
Double valve directing flow to
twin channel (Don McEntee)

FIG. 173
Rocks and stones, which passed
through inlet from spillway,
deposited in channel on right

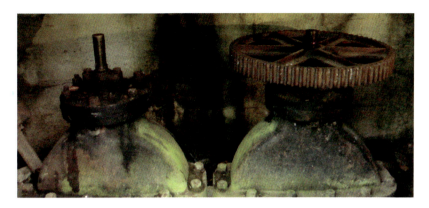

to the second screen wall. A 27-inch pipe connected the outflow to the original
27-inch pipe at "C" [FIG. 108]. The screens were inspected at regular intervals
and the debris removed. This system is still in use today.

Rarely cordial, relations between the Rathmines Commissioners and the
mill owners deteriorated to a point where Sir Robert Herron, James Chaigneau,
Colvill, Frederick W. Pim and other Upper Dodder mill owners secured an
injunction against the township on 18th July 1889. This setback was brought
to the Court of Appeal, which dismissed the action of the mill owners on 28th
February 1890 on the understanding that the Rathmines Commissioners would
execute some additional work. The mill owners then appealed to the House of
Lords, which reversed the judgement of the Court of Appeal. The commissioners
and mill owners then negotiated, reaching an amicable settlement.

Modifications carried out to the works had not been completed in
accordance with terms of the 1880 Act and some essential extra works
were eventually legalised under Section 44 of the Rathmines and Rathgar
Township Act of 1893. The original mill owners' action and later appeal seems
to have been taken somewhat hastily, the main beneficiaries appearing to
be the lawyers. However, this whole unfortunate hiatus did not detract from
the benefits conferred on Rathmines by the Bohernabreena project. The
1893 Act also laid down, that in the event of further disputes involving the
mill owners, appeals would be handled by the Board of Trade, thus avoiding
legal expenses.

FIG. 174
Twin channel to screen house
(Don McEntee)

FIG. 175
Channel entrance
to screen house
(Don McEntee)

FIG. 176
Screen walls inside screen
house (with 2 inch holes)
to filter out small debris
(Don McEntee)

FIG. 177
Superintendant's house
(Don McEntee)

THE SUPERINTENDENT'S HOUSE

A house [FIG. 177], built with stone, was erected for the Waterworks Superintendent. An interesting feature of the original house is that the back portion was the dwelling and the front portion was the fisherman's hall, where the anglers had lockers for their fishing gear. A notable fishing river, it is recorded by Handcock, writing in the 1880s, that quicklime was used by poachers to kill the trout in the river. The quicklime also killed all other fish including eels, sticklebacks, locheen or gudgeon and minnows. The last mentioned fish, Handcock noted, had been introduced to local ponds around the mid-nineteenth century and in time became ubiquitous in the Dodder.

A large feast, with suitable beverages, was held in the fishermen's hall when, early in the morning, the Rathmines and Rathgar Town Commissioners travelled out in horse-drawn carriages for the annual inspection of the waterworks.

Billy Moore lived in the house until his retirement in 1946; AP (Gus) Murphy occupied it from 1946 to 1954. He was responsible for the installation of an electricity generator at the bottom of the left bank of the upper spillway. The water intake to drive the generator was installed to the left side of the slotted weir. This supplied electricity to the Superintendent's house.

Joe McLoughlin, who was the last to occupy the Superintendent's House, lived there with his family for 31 years until he retired in 1986. When the electricity supply failed, he would scramble along the top of the wall at the side of the spillway and make his way to the slotted weir to remove debris from the inlet to the generator. This usually happened during storm conditions and in the middle of the night. This system was replaced when the ESB supply came to the valley in the 1950s.

A colony of bats lives in the rafters of the house. A bat survey was carried out as part of the preliminary investigations for the spillway upgrade scheme. Six different species of bats were found in the valley.

BALLYBODEN WATER PURIFICATION PLANT

The original Bohernabreena Works was designed to supply three million gallons of water daily to Rathmines. The water was drawn off from the Upper Reservoir at three take-off points in the valve tower through a 16-inch pipe; this discharged into the gauge basin "B" [FIGS. 108, 178 & 179] behind the Superintendent's house. This had two functions: one to reduce the pressure on the 15-inch pipe carrying the clean water from the reservoir to Ballyboden and secondly to control the flow up to a maximum discharge to 3 million gallons per day to Ballyboden. A diagrammatic layout of the pipes from the Upper Reservoir is shown in FIG. 121. In 1884, the township demand was approximately one million gallons per day.

The water was carried through a 15 inch (380mm) cast iron pipe which was laid in the bed of the Lower Reservoir [FIG. 108] and then took a 90 degree turn before the lower dam rising to a relief pit on the avenue. The 15-inch pipe continued on to a service reservoir at Ballyboden, a distance of 4.5 miles (7.2km) [FIG. 106].

When constructing a new watermain, scour valves are positioned at all low points on the main and air valves at high points. In the original design the relief pits acted as air valves on the route to Ballyboden. No scour valve was installed under the lower reservoir. It is probable that the pipe under the reservoir became partially blocked, thereby reducing the flow of water to Ballyboden.

At some stage the pipe under the lower reservoir was replaced with a new 15-inch cast iron watermain laid along the avenue connecting into the original pipe at the relief pit at the lower dam. This pipe was connected directly into the 16-inch pipe coming from the Upper Reservoir. With time, cast iron watermains become encrusted with deposits thereby reducing the cross sectional area and consequently the quantity flowing through a pipe.

In 1935 a new 12-inch pipe was laid parallel to the 15-inch pipe to increase the supply to four million gallons per day through the two pipes. The 12-inch and 15-inch pipes were connected directly to the 16-inch pipe coming from the Upper Reservoir [FIG. 121] In 2006 the sections of the 12 and 15 inch pipes in the avenue were replaced by a 600mm High Density Polyetheline (HDPE) pipe.

When the 12-inch pipe was laid, water from the Piperstown Stream [FIG. 134] was piped directly into the 12-inch watermain flowing to Ballyboden. From time to time a blockage would occur in the channel from the Piperstown Stream but would be noted fairly quickly in the treatment plant at Ballyboden as the pH of the water was altered, affecting the running of the plant.

The water was initially conveyed in the 15 inch pipe to the service reservoir in Ballyboden as shown in FIG. 106. In 1935 the 12 inch pipe was laid parallel

FIG. 178 (RIGHT)
Entry to gauging basin
(below upper dam) at
top of photograph with
overflow at upper left corner
(Don McEntee)

FIG. 179 (BELOW)
Exit from gauging basin
(below upper dam) at top of
photograph (Don McEntee)

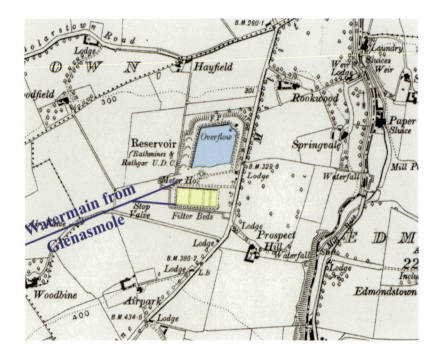

MAP NO. 62

Ballyboden filter beds and
reservoir (OS, 1911)

to the 15 inch pipe. There were two intermediate relief pits along the 4½ mile (7.2km) route en route to Ballyboden. The Water Purification Works were built at Stocking Lane, [MAP NO. 62] where the water flowed into an open service reservoir [FIG. 180] with a capacity of 12 million gallons (54.6 million litres). This reservoir was contained in an earthen embankment at 325 feet (99m) above Ordnance Datum and 175 feet (53.3m) above the highest point in the township, which was at Terenure. The Works was 2¾ miles (4.4km) from Rathmines Town Hall.

The water passed initially through copper wire-gauze screens in the screen house, FIG. 181. When the scheme opened there was no other treatment of the potable water. In 1886 the designers of the Rathmines Waterworks probably felt that the water from Glenasmole (Bohernabreena) was pure enough and did not need treatment. At a later stage the need to improve the quality of water for the township was recognised.

The system of slow sand filtration, to bring raw water up to the standard we expect today, was well established when the waterworks was under construction in the 1880s. A slow sand filtration system was installed as part of Dublin Corporation's Vartry scheme in the 1860s and is the only form of filtration used today at Roundwood. As shown on the 1911 OS Map [MAP NO. 62], a slow sand filtration system [FIG. 182] had been built at Ballyboden to purify the water before it went into the reservoir.

There were five slow sand rectangular filter beds, each of which was 150 feet (46 metres) long. Two were 120 feet (37m) wide, one was 100 feet (30m) and two were 60 feet (18m), giving a total filtering area of 69,000 square feet (6,500 sq. m).

FIG. 180
Ballyboden reservoir
(Don McEntee)

FIG. 181
Screen house,
Ballyboden reservoir
(Don McEntee)

The water from Ballyboden was delivered to Rathmines through a single 18-inch pipe. The township was divided into districts, each governed by a wastewater meter, which proved very useful in localising and checking waste.

No photographs of the Ballyboden plant under construction have so far come to light, but FIG. 183 shows the method of laying cast iron pipes at that time. A simple tripod crane or "gin" was used to lay the pipes into trench. The dress code of the period is noteworthy: engineers and senior management often wore top hats, foremen or supervisors had bowler (hard) hats.

If at any stage the treatment plant at Ballyboden is replaced with a new plant at Bohernabreena the 600mm pipe described earlier could be connected into the potable water network in the Tallaght area.

SUCCESS – AND SOLVING SOME MINOR PROBLEMS

Hassard and Tyrrell's Resident Engineers on the Waterworks Scheme were, successively, Henry Crook and F. P. Dixon. The main contractors were Falkiner and Stanford of Dublin, whose representative on site was E. J. Jackson. Seven hundred men were said to have been employed on the project. Following the successful completion of the scheme, which became fully operational on 11th March 1888, Arthur W. N. Tyrrell M.Inst.C.E. comprehensively described the works in a paper read to the Institution of Mechanical Engineers in London on 25th October of the same year. The paper was titled "Rathmines and Rathgar Township Water Works"

FIG. 182
Slow sand filters,
Ballyboden reservoir
(circa 1900)

FIG. 183
Construction of pipeline
(circa 1860)

Two serious floods occurred during the construction of the Bohernabreena reservoirs, on 1st September 1883 and 16th October 1886. In his paper on Bohernabreena, Tyrrell described the two floods and their effects on the scheme. During the 1883 storm, 3.8 inches (96.5mm) were recorded in a space of nineteen hours. Eight hours after the rainfall started the water topped the embankment of the upper reservoir, flowed over it and carried a large quantity of material to the lower reservoir. The lower embankment, being three feet higher, held for another eight hours before the water broke through the embankment and carried away a large quantity of material. In the 1886 storm, which also lasted about nineteen hours, the rainfall was 3.68 inches (93.5mm), but this flood caused no serious damage to the works.

In the discussion that followed Arthur Tyrrell's paper, it emerged that while 1886 had been very wet, 1887, the year Bohernabreena began supplying water, was the driest recorded in the catchment area up to that time. The drought lasted for 161 days from 23rd May until 31st October, and when it ended, there was still enough water in the Bohernabreena reservoirs to maintain the supply for a further 51 days.

While considering the Rathmines Commissioners' fears of the mill owners, Sir John Hawkshaw noted that the Bohernabreena scheme would provide three million gallons of clear water every day for the Rathmines Township and could cater for between three and four times the then Rathmines population of 30,000 people. The expected daily usage of 30 gallons of pure water per head was in stark contrast to the six polluted gallons per day available to every Dublin citizen in the days of the Poddle supply.

Bohernabreena lessened the worst effects of flooding, exactly as Robert Mallet had envisaged. As with any major engineering work, difficulties that necessitated some deviation from the original plans were encountered during construction of the Waterworks scheme. These were granted legal absolution in

Acts of 1885 and 1892, which also gave some additional powers to the Rathmines and Rathgar Commissioners.

Resulting from the well-deserved success of Bohernabreena, the Rathmines commissioners were able to soften the self-inflicted mortification of the Gallanstown (Grand Canal) works by selling it to Dublin Corporation in 1893. Disposal was authorised under Section 44 of the 1880 Act, other sections of which afforded relief and protection to the Grand Canal Company.

Even after the Waterworks scheme went into operation, some users downstream of the reservoirs continued to take water directly from the Dodder. Frank Cullen discussed water contamination in his *Cleansing Rural Dublin*, which examines public health and housing in South Dublin between 1880 and 1920.

Although the basics of health and hygiene were becoming more widely recognised and practised in the late nineteenth century, the irresponsibility of some unscrupulous individuals and businesses regularly led to court appearances. On 13th April 1891, the water supply to the Carmelite Convent at Firhouse was found to be contaminated. The source was traced to two dairies, which were discharging sewage and other waste to the Dodder. The law had to be invoked to remedy the Firhouse pollution, and several sets of proceedings were taken against other offenders. "Abatement of a nuisance" was how such prosecutions were described at the time.

Pursuant to Section 10 of the 1880 Rathmines and Rathgar Water Act, the Commissioners had to "with all reasonable despatch provide and forever maintain to the satisfaction of the Secretary for War a fire plug at some convenient point within the Township near the North gate of the Portobello Barracks." The barracks, now Cathal Brugha, had access to a supply from the Dublin Corporation Portobello reservoir since the early 1800s.

Section 12 of the Rathmines Act stipulated that "The Commissioners may from time to time establish and maintain such wires and apparatus for the transmission of messages and other communications wholly or partially by means of electricity...for the better execution of any of the powers or authorities for the time being vested in them..." Use of the equipment was strictly confined to waterworks business; making a profit was specifically excluded, protecting as fully as possible the Postmaster-General's position under the Telegraph Act of 1863.

Rathmines took justifiable pride in the Bohernabreena Scheme, which from time to time was the subject of articles in various journals. In the 1908 *British Association Handbook to the Dublin District,* F. P. Dixon, M.Inst.C.E. recorded that Rathmines UDC owned between 50 and 60 acres of property at the lower end of the Bohernabreena gathering grounds. Measures taken against contamination included regular inspection of streams, cleaning catchwater channels and keeping these free of vegetation and debris, and provision of pipe drains separating the farm drainage from the spring water. Beginning in 1900, the council planted nearly 50 acres with about eighty thousand larch, pine, alder and beech trees. The timber was already valuable in 1908, by which time some of the trees had been cut down and used as fencing.

In 1886, when Rathmines had a population of 30,000 people, Bohernabreena was capable of supplying three million gallons (1,363 million litres) of water per day to the Ballyboden reservoir. Ninety-two years later, in 1978, Ballyboden was delivering four million gallons (1,818 million litres) daily and by 1986 this figure had increased again, to six million gallons; however, two millions of this extra water came from the Liffey Works at Ballymore Eustace.

Only about 1.5 million out of the three million gallons that Ballyboden could process daily were needed by Rathmines Urban District Council in 1908. By that year, the population of the Urban District was 39,000 but water was being supplied to a total of 41,000 people, 2,000 of whom were outside the Council's area.

CHAPTER 17

DUBLIN CORPORATION CONTROL – MODERNISATION

n 30th September 1930, a major reorganisation of local government in the Dublin area included, among other important changes, an extension of the municipal boundary. The formerly independent Rathmines and Pembroke Urban District Council areas, together with some parts of the county, now became part of the city. Another momentous feature of the changes was the appointment of Dublin's first City Manager, the distinguished and experienced Gerald Sherlock (1876-1942).

The Bohernabreena Reservoirs and the Ballyboden Waterworks now came under the control of Dublin Corporation, which had previously relied exclusively on the Vartry for the city's water supply. From 1930 onwards, the former Rathmines system became increasingly important as part of the Dublin Corporation network with which it was progressively integrated.

In addition to private residential development, public housing was built on an unprecedented scale during the 1930s. Numerous apartment blocks were erected in the inner city, and extensive housing estates were laid out in suburban areas. To serve all this development, the water and drainage networks were continuously extended and improved, using ever increasing quantities of materials and equipment. As well as Bohernabreena and Roundwood (the Vartry), Dublin was also taking Liffey water from Pollaphuca (Ballymore Eustace) by the 1940s.

World War Two (1939-1945), euphemistically called *The Emergency* in official circles, inflicted escalating universal hardship and deprivation. Essentials such as food, clothing, and fuel were severely rationed and the constantly deteriorating supply crisis called for ever more ingenuity in the public services.

Inventiveness and improvisation became routine, with reclamation and recycling of as much material as possible being practised daily. One memorable instance of this was the removal from Bohernabreena of 24 inch cast iron pipes for re-use in Benburb Street.

During the war there was some—if not great—concern that a stray bomber like that which struck the North Strand area in 1941 might target Bohernabreena. While the classic film *The Dambusters* did not appear until the 1950s, there was a keen sense of what RAF Squadron 617 had done in destroying the Ruhr dams. Michael Murphy recalls that people living below the Lower Reservoir dam ensured that all windows were totally covered by black blinds at night. This despite the luminous intensity of oil lamps and candles being probably invisible to aircraft pilots—the ESB had not yet come to the Glenasmole valley.

Despite materials and money still being scarce in the 1950s, a new 24-inch (600mm) main was laid from Saggart for a distance of 11 miles to bring surplus Liffey water to the Ballyboden (and Stillorgan) reservoirs as required. Now Dublin's three supply systems (Dodder, Vartry and Liffey) were linked.

A drawback of the slow sand filtration system formerly used at Ballyoden was its inability to remove colour from the water. So when some of the peaty water from Bohernabreena got into system, especially in summer time when the river was diverted into the Upper Reservoir, the colour would come through in the water supply. In 1960 a conventional treatment plant of flocculation, rapid sand filters and chlorination was constructed at Ballyboden to treat four million gallons per day (18,000m³/day)—see FIG. 180.

Ballyboden was further upgraded in the 1980s when its sedimentation tanks were converted to upward flow operation and other improvements, including the addition of poly-electrolysis, were carried out. The 18 million litres (four million gallons) per day coming from Bohernabreena were mixed with nine million litres (two million gallons) per day from the Liffey works at Ballymore Eustace to meet the growing requirements of the Ballyboden water supply district. In the mid-1990s a new lime plant introduced the dilution of highly concentrated liquid lime. This reduced the labour element of mixing bagged lime and led to a more uniform lime solution.

A full technical description of the Ballyboden plant and its operation was written by Kevin O'Donnell (Dublin Chief Engineer, Water & Drainage) and Charles O'Connor, Waterworks Senior Engineer, in April 1969 and revised in May 1980. Detailed drawings of Bohernabreena Reservoirs and the Ballyboden treatment works are retained in the Water Service archives.

Dublin City Council is to transfer ownership of the Bohernabreena Waterworks and the Ballyboden Water Treatment Plant together with associated pipelines and works in 2016 to Irish Water.

CHAPTER 18

ℱLOODING ON THE DODDER

 s already explained, the Dodder is notorious for its rapid reaction to rainstorms. The elapsed time (time from the beginning of a rainfall to the peak flow in the river) is often less than 24 hours and, in very severe conditions, can be as little as six hours. Important contributory factors are:

1   The mountainous regions in the southern parts of the catchment which are subject to heavy rainfall;

2   The geology of the upper areas, granite overlaid with peat, which has a high volume of run-off when saturated, possibly as great as 70%;

3   The catchment area, which is large in proportion to the length of the river; and,

4   Ireland's predominant south-west wind

One of the reasons for the construction of the Bohernabreena reservoirs and various ancillary works was to relieve the severe flooding that regularly affected the Dodder and its immediate catchment. Listing and describing such events has therefore been divided into two sections, those that occurred before Bohernabreena was built and those that happened after that time.

FLOODS BEFORE 1880

In his 1652 book Gerard Boate described hazardous rivers or brooks in the Dublin area, and wrote of the Dodder: "Rathfarnum water, of the village by which it passeth... is far the worst of the two, as taking its beginning out of those great mountains southwards from Dublin, from whence after any great rain, such abundance of water is descending to it that the same, which at other times is of very little depth, groweth thereby so deep, and exceeding violent that many persons have lost their lives therein."

On 2nd March 1628—twenty-four years before Boate's book appeared—a witness described a man's death caused by a violent flood in the Dodder: "Carried away by the current, notwithstanding the efforts of his nearest friends and many persons, afoot and on horseback on both sides—nobody was able to succour him." This is eerily similar to the drowning of John Usher in 1629, referred to by Ossary Fitzpatrick.

Acceptably reliable records of the Dodder's history of flash floods, overflowing its banks and causing widespread havoc, extend as far back as the mid-eighteenth century. In the reports of storms that follow, terms such as millilitres or inches of rain refer to the quantity of water falling on an acre (0.404 hectare) of ground. An inch (25.4 millimetres) of rainfall over an acre equates with 100 tons of water, which equals 22,000 gallons of water to the acre. A caveat that must, however, be borne in mind when looking at rainfall figures is that every recorded value defines the precipitation in the immediate area of the recording instrument—there are many instances of heavy rainfall at a particular location while the weather is dry only a short distance away.

In the *Dublin Historical Record* (Vol. 15, 1953), F.E. Dixon wrote that the storm which destroyed Donnybrook and its bridge on 11th September 1739 also flooded the extensive brickfields between Ringsend and Merrion. Faulkner's *Dublin Journal* reported that "Donnybrook Bridge was thrown down....Several ricks of corn and hay, with many chairs, stools and tables etc. were seen floating in Donnybrook river and it is imagined that many cabins have been carried away."

Ringsend suffered seriously in the 1787 flood, Farrar wrote: "It resembled a town which had experienced all the calamities of war, that had been sacked by an enemy, or that had felt the hand of all-devouring time. The unfortunate inhabitants were in a manner excluded from all intercourse with Dublin. They were attacked by the overbearing floods, which issued from the mountains in the irresistible torrents, and completely demolished the bridge."

F.E. Dixon, who was a professional meteorologist and dedicated researcher, believed that the rainstorm of 2nd December 1802 was among the worst ever experienced in Dublin. Three inches of rain (76.2mm) fell in 30 hours, and Ormonde Bridge over the River Liffey was destroyed, as was Ringsend Bridge on the Dodder. Another rainstorm referred to by Dixon occurred in July 1839,

TABLE NO. 1 Major Rainstorms & Floods 1670-1880

1670 June	**1787** 12th November	**1834** November
1719 17th July	**1789** 13th January	**1836** August
1739 11th September	**1792** 4th October	**1846** 21st November
1751 26th February	**1794** 20th November	**1851** January
1782 16th August	**1802** 2nd December	**1877** 3rd January
1784 3rd January	**1807** September	**1880** 26th August
1787 September	**1839** July	**1880** 27th October

when 1.65 inches (41.9mm) of rain fell on Dublin in 24 hours. The January 1851 flood destroyed a wooden bridge which spanned the Dodder at Clonskeagh.

In addition to those described above, there were several other serious floods and storms prior to the construction of the Bohernabreena Reservoirs. Some of the most serious appear in TABLE NO. 1.

The first six months of 1876 were dry, but Dublin's rainfall for the last three months of that year totalled 13.6 inches (345mm), 7.56 inches (192mm) of which fell in December. A newspaper report of 4th January 1877 recorded that "Christmas time was marked by exceptionally heavy rain, but the downpour of yesterday in the city and suburbs was of almost unexampled duration and severity...

"Along the banks of the Dodder, from Terenure to the sea, the fields were inundated to a depth, in some instances, of five or six feet. From Balls-bridge to Irishtown the lands on each side of the road were under water, and the people living in that locality had, if not to vacate the houses entirely, to abandon the lower apartments, which were thoroughly flooded...

"The fields from Sandymount to Lansdowne-road had the appearance of a small sea. The roads in many instances were impassable, and the tram cars had to cease running part of the day."

The establishment of the Bohernabreena reservoirs changed the pattern and gravity of flooding on the Dodder. The flooding occurrences post-1887 are described later. The change of agricultural land to urban developments at various periods along the river and along the tributaries has also changed the flow patterns in the catchment with much quicker runoffs during a storm event. The extent and magnitude of flooded areas changed and the resulting damage increased due to building on flood plains.

Two immutable facts should be borne in mind when studying old newspaper accounts of flooding. First is that many media reports about such events are based on what is said by local people, whose observations are frequently subjective and therefore prone to exaggeration at a time of extreme stress. Second is the tendency, when precedents are being investigated, to ask older people about past inundations Their answers can be affected by the passage of time, hazy recollection and the possibility of exaggeration.

FLOODS SINCE 1887

The former Sewers and Main Drainage Department of Dublin Corporation, now Dublin City Council's Drainage Division, carried out a limited amount of flood prevention and maintenance work on the rivers and streams in its area.

Although the Bohernabreena reservoirs have helped to alleviate flooding problems along the downstream reaches of the Dodder, some serious situations have occurred from time to time. Prevention and mitigation of flooding are in the Drainage Division's remit and reports have been compiled over the years describing various incidents and proposing remedial works. The Dodder's unique and unfailing ability to produce sudden and exceptional runoff rates was constantly watched and the invaluable extra experience gained from each occurrence contributed to the solution of the next flood.

A most important but often overlooked factor in the history of the Dodder—and other rivers in urban areas—was the growth of development along or near these watercourses, creating extensive impervious areas. The rapid surface runoff from such developments can seriously exacerbate flood levels. John Reade, Engineer in Charge in the Sewers and Main Drainage Department in the 1960s, drew attention on several occasions to this increasingly serious problem. In March 1966 he wrote about the Dodder: "Since the beginning of this century there has been considerable encroachment by building development onto the narrow riparian flood plain of the river, so that the previously unimportant and occasional floodings of a riparian field became in time a flood problem involving damage to house property."

Reade calculated that about 3,200 acres (1,295 hectares), or twelve per cent, of the Dodder catchment, was built up at the time of his report. CL Sweeney, surveyor in the Sewers and Main Drainage Department from 1946 to 1988, estimated that the expected flow in a watercourse in a suburban area of the eighteenth century would be approximately half that of the same district fully developed in the 1990s.

In 1977, hydrographic studies of the Bohernabreena Reservoirs and the Dodder were carried out and the results can be consulted in a portfolio of drawings and graphs in the Drainage Division. A report on the feasibility of reservoir control at Bohernabreena was prepared in 1978 by John Clarke of the Sewers and Main Drainage Department. This identified the most critical months for flooding as being from August to December and confirmed the Dodder's peaking time at about six hours.

Listing the vulnerable sections of the river, Clarke identified the section between Donnybrook and Ballsbridge as the most critical at that time. He pointed out that continuing development and increasing property values made Dodder flooding a worsening problem. The very detailed steps necessary to ensure that the reservoir system should be capable of passing a flood with a 10,000-year return period were also set out.

In his report of November 1986, Kevin O'Donnell, Chief Engineer, listed floods on the Dodder since 1880 in order of severity. The chronological list of the eighteen most serious instances over the period 1891 to 2013 appears in TABLE NO. 2.

In the Reservoir Control Report of April 1978, the 1901 and 1915 floods were described as "minor", while those of 1956 and 1957 were classed as "marginal". All the others up to the date of the report were described as "major". The 1931, 1946, 1958 and 1965 storms were reckoned to have return periods of sixty, six, sixteen and thirty-five years, respectively.

TABLE NO. 2 Major Rainstorms & Floods 1891-2013

1891	19th October	**1946**	12th August	**1963**	11th June
1898	23rd November	**1946**	20th September	**1965**	17th November
1901	10th November	**1954**	8th December	**1968**	2nd November
1905	25th August	**1956**	29th September	**1986**	25th August
1915	15th November	**1957**	25th September	**2002**	1st February
1931	3rd September	**1958**	19th December	**2011**	24th October

Reporting on the severe flooding that occurred in the South City on 11th June 1963, John Reade noted that the Meteorological Office classified rainstorms into three categories:

NOTEWORTHY. A storm that occurs once in ten years;
REMARKABLE. Occurring every forty years;
VERY RARE. Occurring every 100 years.

THE AUGUST 1905 AND SEPTEMBER 1931 FLOODS

A major storm, which had serious consequences in the Dodder catchment, occurred on 25th August 1905. The bridge at Castlekelly, adjacent to the Bohernabreena Reservoirs, was destroyed and part of the bypass canal was carried away. Rainfall records for this storm, which also caused severe damage to the Vartry pipeline near Bray, are defective due to a malfunction of the Bohernabreena gauging equipment.

The flood of September 1931 had particularly distressing results. Incessant rain fell for three days, with three inches (76mm) on the third day. Milltown Cottages, long since demolished, were inundated on 4th September when the Dodder flash flood rose 15 feet (4.5m). There was also extensive damage at Lower Dodder Road and Ballsbridge, and the railway bridge at Lansdowne Road was brought down. However, neither 1931 nor any of the other twentieth century storms were as catastrophic as those of 1905 and 1986. Contemporary newspaper accounts of these storms must be set against how similar events were reported eighty years apart.

THE AUGUST 1946 FLOOD

Michael Murphy records that the flood which occurred on 12th August 1946 caused significant damage to the Glenasmole (Bohernabreena) Waterworks infrastructure. To the south of the Upper reservoir, the channel carrying the combined River Dodder, Cot and Slade Brooks was badly damaged. At the bottom of the Upper Spillway the granite wall was breached at the point near where the off take to the screen house is located.

Below the Lower Reservoir, a few hundred yards from the second lodge, FIG. 134, (Michael Murphy's home at that time), where a deep excavation existed on the west side of the roadway, a substantial wall was breached and some of the roadway was washed away. Further down, opposite the White House Farmhouse, the water also breached the river bank and undermined the watermains.

Much other damage was caused along the bed of the river, at least as far down as Fort Bridge. The Bohernabreena Waterworks staff of four persons was augmented by about twenty more to repair the damage, a number staying for several years.

25TH AUGUST 1986: HURRICANE CHARLIE

The summer of 1985 was very wet, followed by more cool and broken weather during the summer of 1986, when soil moisture content was unusually high. An offshoot of Hurricane Charlie moved north eastwards across Ireland on 25th

August, deepening as it did so. It generated prodigious volumes of rainfall, especially in Counties Wicklow and Dublin and parts of the south city area. The resulting flood overtopped the banks of the Dodder in several places, causing widespread hardship to many property owners and economic losses totalling tens of millions of pounds.

In addition to the coincidence of dates, features of the Hurricane Charlie storm are uncannily similar to that which occurred exactly 81 years earlier on 25th August 1905. Both were caused by depressions which tracked along the south coast and up into the Irish Sea.

In the south city, 100mm (4 inches) of rain fell, but at Kippure, it was estimated at between 250mm (9.8 inches) and 285mm (11.2 inches) for the duration of the storm. This rainfall was reckoned to have been the most intense for at least 160 years. The effects of the resultant deluge were made worse by soil saturation in the catchment area because of the wet summer, which led to instant run-off. The flooding that followed was caused by the highest water levels ever recorded on the Dodder: the maximum flow at Orwell Bridge gauging station was between 7600 and 9000 cusecs (215 cubic metres per second and 255 cubic metres per second).

Within the city area, a total of 315 residential and 25 commercial properties suffered damage. The principal exit points for flood waters from the Dodder were catalogued by Jack Keyes, Senior Engineer in the Drainage Division. Thirteen in number, they were at Lower Dodder Road, Dodder Park Road, Orwell Gardens, Dartry Park, Classon's Bridge, Strand Terrace, Clonskeagh Bridge, Beaver Row, Anglesea Road, Bective Rugby Ground, Ballsbridge, Beattie's Avenue and Lansdowne Road. Anglesea Road was identified as the most copious flow exit point.

A first-hand personal account of what conditions were like on the night of 25th August appeared in the *Templeogue Telegraph*: "Having driven from north county Dublin over flooded roads and in atrocious rain, I took the precaution of pulling up and parking my car for a moment as I came towards Austin Clarke (Templeogue) Bridge. I got out of the car and I just could not believe my eyes when I reached the parapet of the bridge. At this place there was—or had been—a fair-sized rock pool in the river below, on the Templeogue Village side. The rushing torrent was now developing a great hole where the pool had been, and the water rushing from this hole was reaching a height of many feet above the level of the road. The noise coming from the river was tremendous. I rushed across to the Tallaght side and found that the water was on the point of breaking out on to the road. It completely filled the eye of the bridge. I did not stick around to see what would happen next. I just took to my heels, got into my car and drove back down Cypress Grove Road."

In the weeks following the storm Corporation personnel were engaged in bringing relief to the victims. Rivers were cleared of obstructions, debris was removed from roads, water pumped from commercial and residential property. House drains were cleared, and houses were completely washed down for old or needy citizens. Jack Keyes headed a team of three young engineers—Anne Graham, Brian Curtis and Neil Kerrigan—who carried out a comprehensive survey. The extent of the flooding was painstakingly recorded, street by street, on a series of ordnance sheets. The large number of maps and other drawings produced at that time showed exactly what occurred and where. These documents have since become the definitive guide for anybody enquiring about flooding in the Dodder catchment.

During the floods of Hurricane Charlie the right hand wall of the upper spillway at Bohernabreena was overtopped, scouring part of the base of the upper dam. FIG. 184 demonstrates the magnitude of the flood coming down the spillway at the upper dam. FIG. 185 gives an indication of the damage caused to the spillway. The breach in the wall and dam (on left of photograph) was repaired immediately after the flood.

It is interesting to note that the staff reading at the lower reservoir in the 1905 storm was approximately 100mm greater than in the Hurricane Charlie storm of 1986, suggesting that the 1905 storm was of greater magnitude. The damage caused by Hurricane Charlie was far more serious than the 1905 storm as large developments had taken place in the flood plain in the intervening years.

An all-embracing report was produced setting out a range of possible remedial works. These included channel improvement by providing a new reservoir or lagoon, raising channel walls, widening the channel and deepening the riverbed. Unfortunately, fully implementing some of these options would have caused other, possibly more serious, problems elsewhere and the cost would have been prohibitive. In his paper *Hurricane Charlie—an Overview*, delivered to the Institution of Engineers of Ireland on 16th November 1987, Jack Keyes described in detail everything that happened during and after the storm, and the remedial work carried out or proposed.

Flooding was also observed on the Poddle (85 properties flooded), the Camac (30 properties flooded) and the Tolka (10 properties flooded) Rivers.

Exhaustive research calculated the return period for a storm of Hurricane Charlie's intensity to be at least 150-200 years, placing it in the chronological range of the one of 2nd December 1802 which, as recorded earlier, discharged three inches (76mm) of rain in thirty hours, demolishing Ringsend Bridge on the Dodder and Ormond Bridge over the Liffey.

Flood defences, such as walls and embankments, are designed to withstand a flood of a particular return period. New flood alleviation works are only implemented if justified on a rate of return dictated by the relevant Government Department. No matter what flood alleviation works are put in place, based on a flood return period, there is always the risk that this will be exceeded at some future date.

THE FLOOD OF 1ST FEBRUARY 2002

The flood that occurred on 1st February 2002 was unusual in that it was not caused by heavy rain or a flash flood in the Dodder—on this occasion, the river rose dangerously because of the tides. The weather forecast for the period from midnight to 18:00 hours (6.00 pm) on 1st February stated: "Stormy conditions are expected overnight and tomorrow; southerly winds with mean speeds of 30-40 mph (48-64 kph) will gust 60-80 mph (96-128 kph). Rainfall up to 30mm will

FIG. 184
Upper spillway in spate during
Hurricane Charlie, 1986

FIG. 185
Flood damage to upper spillway
(Don McEntee)

give some localised flooding." While rainfall was not a significant contributor, key factors were a centre of low pressure, 937 millibars (located over County Down) and the southerly winds. Based on the weather report, Drainage crews were alerted to monitor areas either prone to flooding or possibly vulnerable in extreme circumstances.

The predicted astro tide level for high tide at 14:00 hours (2 pm) on 1st February was 1.93m (Malin Head). What actually occurred was that the tide level reached 2.91m (13:30 hrs), and then dropped to 2.69m at 14:15 hrs, before rising to 2.95m (14.30 hours). The tide statistics in Dublin Port show that the previous highest tide level recorded was 2.59m and this occurred in 1924. Furthermore, the fact that the tide rose and fell twice in such a short period is unusual and noteworthy.

More than 1,300 houses citywide were flooded and the Emergency Plan for Dublin City was activated. This provides a co-ordinated framework within which the major agencies of the State can respond to significant emergencies.

On the Dodder, the wall at Fitzwilliam Quay was breached and flood waters travelled upstream as far as Lansdowne Road. At the embankment and retaining wall built to a level of 3.15 metres northwards from Londonbridge in 1995, water reached to within 50mm (just two inches) of going over the top. In Ringsend, the receding tide caused the collapse of 45 metres of the River Dodder Wall adjacent to Stella Gardens. At 6.15 pm, 35 metres of the quay wall at Stella Gardens collapsed and ten minutes later, at 6.25 pm, another 10 metres disappeared into the Dodder Estuary. This took place in total darkness, relieved by only a few hand torches.

The majority of the houses in this area are single storey, so the residents suffered harshly as so much personal property was destroyed. A further difficulty presented itself when the floods began to recede and evacuated residents wanted to return to their homes. Structural damage seldom occurs in such circumstances and most people think that their houses only need to be cleaned out. Flooding, however, necessitates thorough drying out, rewiring, replastering, new floors and all new electrical appliances. At least six months elapse before a sense of normality returns, and the fear of a recurrence always remains. Some flooded houses were still not occupied a year after the event. The people affected also worry about future flood insurance cover and house values.

Following this flood, which was so very different in many ways from others that preceded it, the City Council formulated a Coastal Zone Risk Assessment Study. The impact of the tidal surge was assessed and the adequacy of the sea defences around Dublin and Fingal was reviewed. An early warning system was devised, based on the results of a new computer model of the sea off Dublin. This type of model has been used in Britain and Continental Europe for some years and the arrangements proposed for Dublin will be progressively integrated into those models.

The flooding of February 2002 was the subject of a paper delivered to the Institution of Engineers of Ireland on 10th March 2003 by Gerard O'Connell and Victor Coe. They described in detail all the main areas in Dublin City which were flooded during the event. These two engineers postulated that the peak tide level was dependent on Astronomical tide levels + an allowance for atmospheric pressure + wind direction and speed + river flows out of the estuary and they developed the O'Connell-Coe formula to predict the maximum high tide level up to four days in advance for the Liffey estuary at Alexandra Basin. It has been in use since March 2002 and the predicted results of storm events have been within 50mm of actual high tide levels. A Triton computer model of Dublin Bay was later developed which also gives reliable high tide levels at many locations around the Bay and Fingal up to 36 hours in advance of an event.

THE FLOOD OF 24TH OCTOBER 2011

Growing interest in climate change is making some of the hitherto obscure or unknown terms now being applied to weather phenomena better known. Occurrences like that of 24th October 2011 are called monster rain and the terms fluvial and pluvial are being increasingly used. Fluvial flooding is that normally connected with rivers. When precipitation exceeds the capacity of the drainage system, the resulting flood is known as pluvial flooding.

Just after 10 pm on Sunday 23rd October 2011 Met Eireann issued a flood warning stating that periods of heavy rain were expected with accumulations of 40 to 70mm leading to flooding; the heaviest rainfall was expected in the eastern coastal counties. The predicted rainfall was not that unusual at this time of the year.

The rain considerably exceeded the forecast, being recorded at the Civic Offices at 95mm over the 24-hour period to 10 pm on Monday 24th. In addition 40mm was recorded up to 4 pm with the remaining 54mm falling over a five-hour period, exceeding the quantity set out in the Met Eireann severe weather alert. The fall was the equivalent of a 100-year rainstorm. No tidal flooding was forecast or experienced. Rainfall of 60 to 90 mm over a four to six hour period is a rare occurrence.

Dublin City, already on alert, and with its resources mobilised, activated the major emergency plan at 8.15pm in the city and South Dublin County Council areas. A Drainage monitoring and assessment team continuously reviewed the latest information, additional personnel being called in as required.

Widespread severe flooding occurred in the city area, the Dodder being a major contributing factor. In the twenty-four hours starting at midday on 24th October, approximately 900 calls for help were received. Some 4,500 sandbags were made available in the Clontarf area. Two thousand were available at Sandymount and 1,000 were delivered to Bath Avenue around 9 pm. Staff members delivering sandbags were delayed by impassable roads and heavy traffic.

About 100 Civil Defence personnel were mobilised and, among many other tasks, arranged emergency accommodation for one family of five and two individuals from Kilmainham and Inchicore. Two families were rescued from marooned vehicles on Wolfe Tone Quay. They were reunited with their families and their cars were towed to secure storage.

Dublin Fire Brigade personnel trained in swift water rescue were deployed during this emergency. Twenty-nine officers and 140 fire fighters were on duty throughout the period; all 21 DFB pumps and other specialised vehicles were in use. The Fire Brigade dealt with a total of 661 calls.

Traffic and transport were severely disrupted. Movement was restricted due to local ponding of water and disabled or abandoned vehicles. The railway bridge at Lansdowne Road (No. 33 in the list set out earlier) suffered damage resulting in the withdrawal of rail services for nearly two weeks while repairs were carried out.

Severe flooding also occurred in the areas of South Dublin and Dun Laoghaire Rathdown, details of which are recorded in the offices of both county councils.

In December 2011 it was estimated that 192 dwellings and 136 other buildings/non-residential ground floors flooded during this event. Several sports grounds were affected and more than 200 vehicles were destroyed.

Apart from the Dodder and Poddle, five other rivers and several districts throughout the city suffered. Up to the end of December the number of buildings reported flooded throughout the city totalled 1,250 with 750 related to river/estuary flooding.

After every flooding, a thorough examination of all the factors involved in the event is carried out. Appropriate plans are then prepared to alleviate the likely effects of similar future occurrences.

FLOW RECORDERS IN DODDER CATCHMENT

The very useful Orwell Bridge recorder, dating from 1952, was upgraded in 2000 and is now maintained by the Environmental Protection Agency (EPA). Dublin City Council installed a recorder on the Dodder at Bohernabreena in 2005 as part of the Spillway Upgrade project. In December 2011 this recorder was connected to the City Council's telemetry system. The real time flows are now sent to the City Council's monitoring system and the EPA's hydrometrical section. There are also stations on the Owendoher (Willbrook Road) and the Slang River at Frankfort. A new river/tidal gauge, which was installed at Beatty's Avenue in Ballsbridge in 2008, was washed away in the flooding of 24th October 2011.

A rain gauge, with a 24-hour chart recorder, has been continuously recording the rainfall at Bohernabreena for over 100 years. There are also a number of daily rain gauges in the catchment.

CHAPTER 19

FLOODING ON THE DODDER TRIBUTARIES

FLOODING ON THE PODDLE

While floods are usually thought of in relation to the larger rivers, the smaller ones can also create dire local problems, as the Poddle did at Lower Kimmage Road and Harold's Cross in the 1980s. Because it is a minor tributary of the Liffey, it is sometimes not realised that the Poddle is tidal. If an unusually high tide coincides with heavy rain and an east wind in Dublin Bay or the Liffey, the Poddle is capable of causing serious trouble. A constant fear is the unpredictability of conditions severe enough to produce a storm with a return period so long as not to have been previously recorded.

A report of an occurrence that took place in 1814 illustrates how a combination of a river flood and tidal conditions can interact to produce what, fortunately in this instance, had a happy outcome. On this occasion, when the Poddle was in spate, a boy who was on an arch being built in the Castle Yard fell into the river and was carried down under Dame Street and Wellington Quay to the Liffey, from which he was fortunately rescued because it was low tide.

Flooding which occurred on the Poddle during Hurricane Charlie in August 1986 was described in Jack Keyes's comprehensive report of December 1986 as among the worst on this river since records began. The most severe flooding occurred at Kimmage Cross Roads, Ravensdale Estate, Mount Argus and Harold's Cross.

Eighty residential and five commercial properties were affected, causing great disturbance and hardship to local people. The Corporation carried out an immediate clean-up, as much help as possible being given to those most in need. All the places and properties affected were examined in minute detail and appropriate remedial works were recommended.

At Lower Kimmage Road, the Poddle culvert passes under the front gardens of more than twenty houses. In 1985, a lengthy section of this culvert collapsed and had to be replaced. Jack Keyes noted that only 15% of the Poddle's course had sufficient capacity to deal with floods but this would improve when the front garden culvert along Lower Kimmage Road was renewed. This was constructed in 1988-1989 with Gerard O'Connell as resident engineer.

Echoing John Reade's warning of twenty years earlier about the increased risk of flooding in the Dodder as development in its catchment intensified, Keyes now expressed similar concerns for the Poddle. As had been done for the Dodder, maps and drawings were produced illustrating Poddle flooding and detailed proposals to alleviate problems in the future.

Severe flooding took place on the Poddle on 24th October, 2011. River screens were blocked with debris carried down during the flood at the Lakelands overflow. The river burst its banks allowing the flow to make its way overland. The Gandon Hall screen (near Harold's Cross) in turn became blocked by debris carried down in the flood, causing extensive inundations downstream. There was, regrettably, one fatality.

Eleven locations and 113 premises were affected and there was considerable damage to vehicles, most of which were write-offs.

FLOODING ON THE SLANG RIVER

The Slang (Dundrum) River has a surprising history of flooding. Before the housing and industrial development occurred in its catchment flooding did not cause any extensive damage to property. The most notorious events are given below in historical sequence.

FLOOD OF 1936. A major flood came down the Slang in 1936 inundating the forty cottages in the valley at Windy Arbour. The force of the flood waters demolished some of the houses and rendered the rest of them uninhabitable. The houses were all demolished after the flood and many of the families lived in Dundrum Library for some months until they were rehoused.

FLOOD OF 1963. The city's worst flooding since 1954 occurred on 11th and 12th June 1963. One of the worst electrical storms in living memory left a trail of damage in its wake after it raged over Dublin for more than three hours in the afternoon and again in the evening. The storm was centred over Mount Merrion where over 7.25 inches of rain fell on 11th June with 3.25 inches falling between 1350h. and 1455h. in the afternoon. This resulted in a wave of water cascading down from Mount Merrion, along Fosters Avenue and overland to the sea at Merrion. Walls were washed away and torrents of flood water passed through many houses and flooded others to a depth of 4 feet.

A shocking tragedy occurred at the junction of Fenian Street and Holles Street when an already weakened Georgian building collapsed. This had a shop on the ground floor and two children who were buying sweets were killed.

There was extensive flooding in Dundrum with Rosemount the worst hit area. A large flow came down the abandoned railway line flooding the crossroads at Churchtown Road and Dundrum Road. Flooding also occurred in Wyckham and Meadowbrook estates adjacent to the Slang River. Most of the roads in the south county were blocked by flood waters.

THE NOVEMBER 1982 FLOOD. A major flood occurred on 5th, 6th and 7th November 1982. The Slang River overflowed its banks on Ballinteer Road and in the Pye lands and the flood waters followed the old natural river channel down to the Pye centre in Dundrum. All the small enterprises in the centre were flooded as were the basement area and the bowling alley at ground floor level. The basement was used as a printing works at the time and the flood waters rose so quickly that two of the printers were trapped. They managed to squeeze out through a window just before the basement was inundated.

HURRICANE CHARLIE. Hurricane Charlie (25th August 1986), already described, caused extensive surface water flooding in the Dundrum area. The major flood damage during this storm occurred along the Dodder River.

MAY 1993. In May 1993 the Pye factory at Ballinteer Road was again flooded from the Slang River

OCTOBER 2011. The flood of 24th October 2011 has already been described in some detail. When the Dundrum Town Centre (built on Pye lands) was constructed a culvert was built through the underground car park to take the Slang river from an open channel on the east side to another open channel on the west side. The entrance to the culvert became blocked with debris caused by flood waters. The Slang River flowed over the sides of the culvert and flooded the Dundrum Town Centre causing major damage to the ground floor shops and restaurants. The owner of a Mexican restaurant in the complex said that five feet of water had rushed down steps towards his business, causing thousands of euro worth of damage. The ground floor apartments upstream of the Centre were also flooded.

There was also extensive flooding in several other areas in South Dublin.

CHAPTER 20

DROUGHTS AND SNOWFALLS

DROUGHTS

We have seen how floods pose a serious threat to peoples' lives, homes and workplaces, which sometimes have to be vacated in the worst possible circumstances. Simultaneously, serious damage is inflicted on bridges and other structures, attracting maximum attention and detailed reports. Because droughts, in contrast, rarely threaten life and buildings, they tend to slip in and out of public awareness and are recorded in far less detail than floods.

The difficulties created by droughts are often related to conditions prevailing at particular periods. For example, Robert Mallet described how, in the 1840s, a lack of water could bring industry along the Dodder to a standstill. In subsequent years, as steam, gas and electricity superseded water in energising industry, the number of riverside mills declined steadily, but those that survived were at least assured of a constant supply of water following completion of the Bohernabreena Reservoirs in the 1880s.

Long before the environment became a subject of widespread interest, most people regarded a drought as a minor downside of a fine summer. In Dublin, this relaxed attitude was sharply challenged in 1893, when the much-appreciated but by then taken for granted Vartry water supply failed for the only time in its history. Even if it was not understood at the time, this was an early lesson in environmental awareness.

During the twentieth century, as science progressively enhanced the quality of life, increasingly forensic studies were carried out on every conceivable aspect of weather and the seasons. Drought inevitably came under sustained scrutiny which has become ever more important with the growing attention to climate change, global warming and the environment.

The classification "Absolute Drought" tops Met Eireann's table of drought severity, describing a period of fifteen or more days on none of which there was 0.2mm or more of rain. Using this definition and the concept of a dry year, it is worth recording that, contrary to widespread beliefs, droughts can occur in winter as well as summer.

In "An Assessment of the 1995 Drought" the Environmental Protection Agency (EPA) examined droughts that occurred in the last quarter of the twentieth century. The years 1976, 1989, 1990, 1991 and 1995 were classified as dry years, the low river flows of 1976 being taken as the standard against which other severe droughts should be measured.

Unfortunately, the definition and methodology used to describe droughts in the late nineteenth century were different to those employed today. It is recorded that 1887, the year in which the Rathmines water supply system came into operation, was the driest recorded in the Dodder catchment area up to that time; this drought lasted for 161 days from 23rd May until 31st October.

To allay the fears of environmentalists, the frequency of low water flows in the Dodder will need constant monitoring. Robert Mallet noted that some river catchments respond rapidly to rainfall and are regarded as flashy, with little or negligible storage—a description that applies very accurately to the Dodder.

There are several historical references to low water flows or droughts in the capricious Dodder, many of them recorded as events rather than scientific observations. Today, a society more attuned to the implications of drought for climate change and global warming looks on all such happenings with increasing interest. The frequency and scale of low water flows in the Dodder are being recorded and analysed to yield useful information in planning how to manage river water as an invaluable resource.

SNOWFALLS

The most dramatic opposite to the drought, in Ireland at any rate, is snowfalls which in recent years has been relatively rare. When a heavy snowfall occurs and freezing temperatures continue for some time, the effect on rivers can be delayed until the thaw sets in. Flooding can then occur.

Snowfalls in the mountains have caused serious problems for local residents and Waterworks staff, especially in terms of personal safety and accessibility, as well as the provision of essential supplies. There are records of heavy snowfalls and blizzards back over many years; some of the most occurrences are given here.

A very heavy snowfall struck Dublin on 1st January 1786, depositing three feet of snow. A blizzard which hit the country on 19th and 20th November 1807 caused the deaths of several people who were travelling at the time. What was possibly the worst snowstorm of all, in January 1814, rendered roads impassable for months and £10,000 was raised to assist 66,000 people who were badly affected by the miserable conditions. This particular snowstorm resulted in arrangements being made for "snow, and every obstruction caused by frost, to be scraped off the steps of doors, and from the footways and paved gutters or channels running along houses and premises; and coal-ashes, sawdust or sand, to be strewn on such footways during frost". Who was to be responsible for this Work is not recorded.

There were further snowfalls and blizzards during the nineteenth century with an especially tragic happening on 14th February 1853 when the Queen Victoria was wrecked off Howth with the loss of 55 lives.

Moving on to the twentieth century, a serious snowstorm occurred in 1933 The blizzards of February and March 1847 were the most serious experienced in Ireland for several years. Many areas were cut off for weeks by huge snowfalls. Bohernabreena was isolated and the blanket of snow was so deep that roads had to be located before clearing could begin. At that time, Ireland was a poor country lacking financial and material resources and snowploughs were not available. John Lee recorded that as soon as men and youths using shovels cleared the roads, fresh snowfalls and drifting snow covered them up again. The ESB received help from the Defence Forces whose off-road vehicles such as artillery tractors, could reach otherwise inaccessible fallen cross-country power lines. Conditions did not improve until after St Patrick's Day and there was still snow on the mountains in May. Fortunately, however, an excellent summer followed.

There were further serious snowfalls in January 1963 and yet another very severe period of blizzards on 8th January 1982. For weeks after that, frozen snow was so hard packed on several main roads that JCBs had to be used to load it into waiting lorries for disposal. Traffic and essential public services were disrupted for more than two weeks and water levels in the Dodder — and other rivers — were much higher than normal as a thaw set in.

Further particularly bad and persistent periods of snow occurred in the last weeks of 2010 and the early days of 2011 and again at the end of 2011.

CHAPTER 21

BOHERNABREENA SPILLWAYS UPGRADE

RESERVOIR SAFETY

The Reservoir (Safety Provisions) Act of 1930 (UK) and—much later—the Reservoir Act of 1975 (UK) placed responsibility for safety on the owners of reservoirs with an impounding capacity of more than 25,000 cubic metres. In the public interest, reservoirs are to be inspected every ten years by a competent Panel Engineer.

At the request of Dublin Corporation, specialist Consulting Engineers Binnie & Partners carried out inspections at Bohernabreena in December 1974 and March 1975 in line with the regulations in the United Kingdom.

Binnies reported that the appearance of both embankments was good, with no evidence of slipping on the downstream slopes of either dam. While the general condition of the installations was satisfactory at the time, detailed hydrological recommendations were made to improve the efficacy of the spillways and dams. In the light of a further guide to reservoir safety, Binnies again examined the Bohernabreena dams in September 1986 and additional improvements were recommended. The major improvements were the upgrading of the spillways to allow the safe passage of the Probable Maximum Flood (PMF).

As there was no provision in law in the Republic of Ireland for the construction and operation of reservoirs, Dublin City Council appointed Andrew Rowland of Binnie & Partners (later Binnie, Black and Veatch) as All Reservoirs Panel Engineer under the United Kingdom 1975 Reservoirs Act to advise Dublin City Council on the design and construction of the entire spillway works.

FLOOD DANGERS AND REMEDIES

As noted, above, Binnie & Partners, Dublin Corporation's Panel Engineers, carried out an examination of the dams after the storm of 1986. During the storm, the lower dam was within inches of being overtopped. The flow from the upper reservoir overtopped the right hand (east) wall of the spillway, FIG. 185, and caused small undermining of the downstream slope of the upper dam embankment

It is interesting to note that during a storm at Rotherham in England in June 2007, when 90mm of rain fell in 18 hours, undermining of the downstream slope of the Ulley dam embankment occurred similar to that at Bohernabreena. With the Ulley reservoir full, a torrent of water spilled out, causing significant structural damage to the masonry channel walls and extensive damage to the dam itself. The council acted quickly; the M1 motorway was closed and 1,000 people in the local villages were evacuated for two days while the dam was being being repaired.

Following their detailed study of the dams and associated works in 1987, Binnie & Partners concluded that the dams at Bohernabreena were in danger of being overtopped. As these dams are of the earthen embankment type a breach of either would lead to major property damage and almost certain loss of life downstream and they thus posed a serious threat to the population resident in the lower reaches of the River Dodder in urban Dublin.

As neither of the spillways at the Bohernabreena dams was capable of passing the Probable Maximum Flood (PMF), Binnies strongly recommended urgent and complete reconstruction. In the interests of public safety, the reconstructed spillways should be capable of catering for Probable Maximum Flood (PMF) from a one in 10,000 year storm.

THE SPILLWAYS PROJECT – DESIGN

Dublin Corporation engaged the services of Binnie, Black & Veatch (BBV), Consulting Engineers to advise on the upgrade of the spillways. Model studies were carried out at University College Dublin in tandem with hydraulic analysis by BBV. The initial design was undertaken by the Design Division of Dublin City Council with Gerard O'Connell as Project Engineer based on physical 3D models built under the supervision of Aodh Dowley in UCD.

At this stage the lower spillway was to be reconstructed first, followed by the upper spillway. Ancillary works would be undertaken throughout the construction period. Access to the lower spillway was along the existing road through the waterworks. Access to the existing upper part of the waterworks and construction works of the upper spillway was planned from Upper Castlekelly Road. The Planning Authority, south Dublin County Council, did not approve the initial design concept as they would not permit the Upper Castlekelly Road to be used by construction traffic accessing the upper spillway and works.

Under the guidance of Don McEntee, Senior Engineer, the design team of Noel Murphy, Bill Jamison and Terry O'Neill redesigned the scheme to comply with the requirements of the planning authority. The access to the works had to be maintained along the existing road. This necessitated a small modification to the proposed layout of the lower spillway and the construction of a piled wall to keep the road open at all times. Planning permission for the project was obtained from South Dublin County Council in 1997. Binnie Black and Veatch were then commissioned to prepare the contract drawings and documents in conjunction with City Council engineers.

The tendering process and tender evaluation were carried out by Dublin City Council, Design Division (Water). The preferred contractor was Charles Brand-Bilfinger Berger, Joint Venture (CBBBJV). Construction commenced on 22nd January 2002 and was completed on 22nd December 2005. Sean Donnelly was the resident engineer on the project. Reconstruction of the spillways together with upgrade to the waterworks took four years and cost €13 million.

SPILLWAYS CONSTRUCTION DETAILS

As the Upper Reservoir was the source of the water supply to Ballyboden treatment plant with an output of 18 million litres per day, a sequence of works was set out to ensure that this supply was maintained without interruption throughout the duration of the works. A key requirement of the scheme was that the contractor had to provide for the management of floods during the course of the works. The contractor had to design the works sequence so that the 1 in 10 years return period flood could be passed without damage to the works in progress and the 1 in 100 year return period flood could be passed without any risk to the dam. The main restriction on a severe flood was the existing two-arch bridge across the downstream end of the spilling or tumbling basin to the lower reservoir. Demolition of this bridge early in the construction sequence allowed the 1 in 10 year flood to pass down one half of the channel and ensured that the 1 in 100 year flood could pass safely during construction.

The planning authority (South Dublin County Council) wanted the original masonry appearance of the new spillway to be replicated. A reinforced concrete wall with masonry facings would have been impractical in a spillway setting with huge scouring forces at play in a major storm event. To meet the requirements of the planning authority the design was revised to a textured finish to the concrete spillway walls giving the appearance of dressed masonry and similarly an imprint finish to the base slab to mimic the impression of random masonry.

The reservoir levels were also to be kept at a maximum of Top Water level (TWL) minus 4m for the Lower reservoir and TWL minus 2m for the Upper reservoir during the duration of the works.

LOWER RESERVOIR

At the lower reservoir the weir and twin span bridge shown in FIG. 186 had to be replaced with a reconstructed weir discharging into a deeper stilling basin and a single span bridge [FIG. 187]. Debris tended to get caught in the central pier thereby impeding the flow and increasing the water level in the lower reservoir above the weir level. A new 1.6 metre diameter valve was incorporated at the upper end of the lower reservoir spillway stilling basin approximately 4m below the weir crest level [FIG. 128]. This provided a greater drawdown rate from the lower reservoir in the event of a major storm and reduced the peak flow duration.

One of the big problems to be faced during construction was how to protect the dams in the course of construction if a major flood event occurred, as the new spillways were being built in the same location as the existing ones. The contractor had to provide temporary works to allow the 1 in 10 year flood pass without damage to the spillway construction works and the 1 in 100 year flood to pass without damage to the dam.

The scheme involved the removal of the existing masonry spillways and their replacement with reinforced concrete spillways. The new lower spillway has a capacity of the order of 380m³/S, three times the previous volume.

Other elements of the scheme included a contiguous piled wall to the west (left side) of the Lower Reservoir stilling basin, heightened wave walls to both dams, a new 600mm diameter water main, improved access road and a car park at the entrance.

The demolition and reconstruction of the lower spillway was done in two phases with the left (looking downstream) side rebuilt first. A critical element of this first phase was the construction of a contiguous piled wall (designed by Arup Consulting Engineers) on the left abutment to the stilling basin which effectively retained the adjacent steep hillside when the deep excavation of the stilling basin was taking place and permitted the access road to remain open. FIG. 128 shows the reconstructed lower weir with the new 1600mm valve at the upper end. FIGS. 188, 189, 190 & 191 show the weir and spillway under flood conditions.

FIG. 186

Old weir and double span bridge to lower reservoir (Don McEntee)

FIG. 187

New weir, stilling basin and replacement bridge to lower dam (Don McEntee)

FLOOD EVENTS

In the course of construction of the lower spillway two significant floods occurred on 31st October 2003 and 2nd December 2003. The December event resulted in a breach of the temporary cofferdam and damage to the west end of the dam. The puddle clay core of the dam remained intact. The repairs to part of the original dam were undertaken in a methodical fashion in 150mm layers until full reinstatement was completed.

UPPER RESERVOIR

The bywash channel and weir were incorporated into one stilling basin. This necessitated the removal of the bridge shown in FIG. 146 and the construction of the two bridges as shown in FIGS. 192 & 193. This was to keep access to the upper dam and the avenue between the bywash channel and the reservoir as shown in FIG. 194.

The construction methods used in the lower spillway works were replicated in the upper spillway. One major difference was that the design of the upper spillway included the provision of two splitter walls running along the centre of the spillway [FIG. 195] creating three channels in the spillway. This was to prevent cross waves overtopping the side wall and thus scouring the base of the upper dam. The construction of the left splitter wall on the west side then obviated the need for a temporary wall during construction of the east side of the spillway.

FIG. 188
Flood flow over reconstructed
lower weir and stilling basin
(Ger Goodwin, 2008)

FIG. 189
Flood flow over
reconstructed lower weir
(Ger Goodwin, 2008)

FIG. 190
Flood flow in reconstructed
lower spillway (Ger
Goodwin, 2008)

FIG. 191
Flood flow in reconstructed
lower spillway at footbridge
with flow from 27 inch pipe on
left (Ger Goodwin, 2008)

FIG. 192
Replacement bridge over
bywash channel to upper
access road (Don McEntee)

FIG. 193
Replacement bridge over
upper spillway to upper dam
(Don McEntee)

FIG. 194
Upper reservoir access road
with reservoir on left and
bywash channel on right

FIG. 195
Reconstructed upper spillway
with splitter walls (Don McEntee)

FIG. 196
Reconstructed upper weir,
stilling basin and bridge
over bywash channel
(Don McEntee)

FIG. 197
Flood flow over reconstructed
upper weir on left and from
bywash channel on right
(Ger Goodwin, 2008)

The new weir, stilling basin and bridge [FIG. 196] replaced the slotted weir, two bridges and weir. FIGS. 197 & 198 show the weir and spillway in flood conditions.

The access road from the entrance gate to the Superintendent's Lodge was upgraded during the course of the works. However, the original access road from the Superintendent's Lodge to the Upper Reservoir was not suitable for construction traffic—so an alternative temporary road was constructed west of the original road to enable the Upper Dam works to proceed.

POST-CONSTRUCTION INSPECTION

Following inspections after completion of the works the Panel Engineer, Mr. Rowland, certified that the recommended works had been implemented. He also set out recommendations for the safe operation, maintenance, monitoring and surveillance of the dams into the future.

FIG. 198

Flood flow in upper spillway
(Ger Goodwin, 2008)

CHAPTER 22

CATCHMENT–BASED FLOOD RISK ASSESSMENT AND MANAGEMENT STUDIES (CFRAMS)

ollowing the major flooding that occurred in Ireland in the first decade of the twenty-first century the Office of Public Works was designated as the authority responsible for implementing the EU Flood Directive in Ireland. To assess the risks and management of flood prone areas they have to complete Catchment-based Flood Risk Assessment and Management Studies and Catchment Flood Risk Management Plans (CFRAMP) for the country by December 2015; these are known as CFRAMS and CFRAMP.

The objectives of a CFRAMS & CFRAMP are:

- Identify and map existing and potential future flood hazard risk within the catchment. Identify viable structural and non-structural measures and options for managing the flood risk.

- Build a strategic information base necessary for making informed decisions in relating to managing flood risk.

- Develop environmentally, socially and economically appropriate long term strategy (Catchment Flood Risk Management Plans) to manage the flood risk and help improve the safety and sustainability of communities in the catchment up to the National standards if viable.

- Carry out a Strategic Environmental Assessment (SEA) and other appropriate assessments to ensure that environmental issues and opportunities for enhancement are considered.

The requirements of the EU Directive on the assessment and management of flood risks (2007/60/EC –the Floods Directive) were transposed into Irish law as Statutory Instrument 122 of 2010.

DODDER CFRAMS

The Dodder CFRAMS is one of four OPW pilot studies in Ireland and is the first comprehensive study undertaken with a view to producing a single flood risk management strategy for the whole Dodder catchment. Consultants RPS were appointed to carry out this study in 2007.

The key outputs from the Study are flood hazard and risk maps, a Catchment Flood Risk Management Plan (CFRMP), Strategic Environmental Assessment and Appropriate Assessment. The main aim of the Study is to undertake a comprehensive flood risk assessment of the Dodder River catchment. The study includes the development of a robust computer model representing the hydrological and hydraulic characteristics of the River Dodder catchment. This model was used for the mapping of the flood risk.

Using model results, the environmental, social, technical and economic merit of various options for flood management, taking account of current and future land developments, is presented in the final report. Consideration of the latest findings in relation to the effect of climate change on relevant issues such as rainfall, tide levels and river flows was also undertaken. As part of the flood risk management plan a River Dodder Maintenance Plan was prepared for the river and its main tributaries. The draft final plan and draft SEA and Appropriate Assessment were put on display from March to June 2012. The CFRAMP was adopted by Dublin City Council in March 2013 and by South Dublin County Council in April 2014. It was noted by Dun Laoghaire Rathdown County Council in November 2014.

STRATEGIC ENVIRONMENTAL ASSESSMENT (SEA)

FLOOD RISK MANAGEMENT PLAN. In accordance with the recently introduced EU regulations a Strategic Environmental Assessment (SEA) will be undertaken as part of the study running in parallel with the development of the Flood Risk Management Plan. The objectives of the SEA are to consider the environmental constraints, and opportunities, within the catchment, and to look at the environmental consequences of choosing one option relative to the impact of choosing a reasonable alternative option, or options, at a strategic level in order to minimise any environmental impacts. An integral part of the SEA process is consultation with the public and all relevant stakeholders. The process will include public information events.

On 24th January 2008, a stakeholders' workshop to consider reports on the Dodder and identify future objectives was held at the Civic Offices, Wood Quay. Organised by Don McEntee and Gerry O'Connell of the Projects Division, it consisted of various presentations, discussion by eight task groups whose conclusions were then given to the attendees for further consideration.

The event was opened by Gerry O'Connell of Dublin City Council and the principal presenters included John Martin, OPW Engineering Service, Bjorn

Elsaesser, Claire Coleman and Dr. Marian Coll of RPS Civil Engineering, Dr. Mary Tubridy, Habitat Studies.

SEA stakeholders fall into three categories. The three environmental authorities are:

- Environmental Protection Agency (EPA)
- Department of Environment, Heritage and Local Government (DEHLG)
- Department of Communications, Marine and Natural Resources (DCMNR).

The Primary Stakeholders are:

- Office of Public Works (OPW)
- Dublin City Council (DCC)
- Dublin City Heritage Office
- Dublin City Planning
- South Dublin County Council (SDCC)
- Dun Laoghaire-Rathdown County Council (DLRCC)
- River Basin District – Eastern Region (ERBD)
- Eastern Region Fisheries Board (ERFB), now Inland Fisheries Ireland
- National Parks and Wildlife Council (NPWS)
- The Heritage Council
- Secondary Stakeholders:
- Central Fisheries Board
- Bird Watch Ireland, Dodder Valley Project
- Dublin Transport Office (DTO): Irish Rail/Dublin Bus
- Dodder Anglers Group
- Dublin Naturalists Field Club
- Bat Conservation Ireland
- Geological Survey of Ireland
- An Taisce, the national trust for Ireland
- Waterways Ireland
- Irish Wildlife Trust
- National Roads Authority (NRA)
- ESB
- Sustainable Water Network
- Coillte
- Teagasc
- Marine Institute
- Irish Farmers Association (IFA)

Following the presentations, they divided into eight groups and deliberated on the following subjects:

NO. 1 Biodiversity, Flora and Fauna;

NO. 2 Water quality and quantity;

NO. 3 Landscape and Visual;

NO. 4 Cultural Heritage;

NO. 5 Material Assets and Infrastructure;

NO. 6 Human Impacts;

NO. 7 Fish;

NO. 8 Environment

The workshop concluded with a representative from each group offering a synopsis of what they considered should be included in future plans for the Dodder. It was agreed that further research should be undertaken and meetings held to review progress, ensuring that the agreed improvements are implemented The workshop also concluded that that Human Health or threat to Human Life was the most important criteria. Next most important was failure of critical infrastructure as it affects human life. All other criteria to be assessed on an individual case basis.

FIG. 199
Reconstructed Dodder river walk and new flood protection wall at rear of footpath, Ringsend (Don McEntee)

ALLEVIATING FLOODING

The Dodder Flood Alleviation Scheme (Phase 1) is intended to deal with flood risks in the river's lower catchment. Two types of flood have been identified: (1) a coastal or marine flood, similar to that of 1st February 2002 and (2) fluvial, triggered by heavy rainfall in the river's catchment, similar to October 2011. Coastal floods occur downstream of Londonbridge, while fluvial flooding, which can result in higher water levels, happen upstream.

National design standards are now applied to works carried out in accordance with Office of Public Works (OPW) stipulations. Defences intended to resist coastal flooding should be capable of withstanding a one in two hundred years inundation, while those designed to protect against fluvial torrents should be able to hold out against a flood with a return period of once in 100 years. New works should incorporate freeboard allowances of 300mm for hard and 500mm for soft defences.

Before work began, Phase One of the Dodder flood alleviation programme was carefully planned, taking into account

Existing defences and their condition;

Non-flood defence objects, e.g. trees, service, and buildings;

Environmental, social and archaeological issues;

Methodology

The first part of the scheme, which included the reconstruction of the floodwalls at Stella Gardens, was completed in 2008 and the section at Derrynane Gardens was also substantially complete by the end of 2008. The section from Londonbridge to Lansdowne Road bridge on the Aviva Stadium side was opened in December 2009 with the section on the opposite side largely opened in May 2010 [FIG. 199]. Construction work on the section at Marian College began in August 2011 and was opened to the public in May 2012. A 35m section of the quay wall at Fitzwilliam Quay beside Ringsend Bridge collapsed on 6th February 2011; reconstruction works started in November 2011.

Works commenced on the section beside the old Sweepstakes site early in 2012 and these were opened to the public early in 2013. Emergency works were carried out in the Licenced Vintners and Anglesea Lane off Anglesea Road in 2013. Works were completed at the Oval Shopping centre, Landsdowne Woods, Ballsbridge Woods and Ballsbridge Gardens Apartments in 2014. Works are currently being constructed on Beatty's Avenue, adjacent to Estate Cottages, in Herbert Park, and the Licenced Vintners. Further works are planned upstream of Ballsbridge to Beaver Road weir, from 2015 to 2016 as per Part 8 approval on Dublin City Council web-site.

Works are also planned in the future in South Dublin and Dun Laoghaire Rathdown following statutory planning procedures.

The flood of 24th October 2011 which inundated 324 buildings in the City area has heightened the requirement for these new flood alleviation works. Over a dozen large flood gates have been installed so far on this project, even with these gates left completely open in flood events tidal or fluvial flood risk is reduced by over 95%. These gates mean that the public and visitors can get close to the estuary and river in non-flood situations. These defences are designed to last for 100 years with a provision for estimated global warming to the year 2100.

Vegetation is starting to grow back in disturbed areas of the Dodder Estuary and wildlife such as Swans, Ducks, Otters, Brent Geese, Herons and sea birds have returned in similar numbers to previous. Fish passes are also planned on various weirs on the river, including the Clonskeagh and Beaver Row weirs to allow Salmon to go up stream to spawn.

PODDLE CFRAMS

Following the severe flooding in the Poddle Catchment in October 2011, Dublin City Council and South Dublin County Council asked the OPW if it was possible to fast-track the CFRAM Study of this river. Appropriate funding was found and the Draft Study and Plan was completed late in 2013 which highlighted three viable options. The Plan went out to public consultation in 2014. Documents are now being compiled to appoint consultants to carry out preliminary design of the preferred of these three options, which involves storage in Tymon Park as well as new flood walls and embankments in both Council areas, and bring it through the Planning Process.

CHAPTER 23

℘OLLUTION AND AMENITY

POLLUTION

Recorded efforts to combat river pollution in the Dublin area go back at least as far as the eighteenth century, when people complained that the public water supply was unfit. The cause was traced to effluents from tucking and paper mills on the Poddle, the waters of which came from the Dodder. To remedy this situation the Irish Parliament enacted legislation in 1719 "for cleaning and repairing the watercourse from the River Dodder to the City of Dublin, and to prevent the diverting and corrupting of the Water therein". Following its cessation as the source of the city's water supply in 1778, the Poddle became an increasingly and notoriously polluted river, mainly from sewage and industry, until the Dublin Main Drainage Scheme came into operation in 1906.

The assorted mills and other riparian industries along its banks poured significant quantities of waste into the Dodder. In Mallet's 1844 report, the list of Dodder industries included at least three paper mills. Paper-making is widely regarded as being among the most serious contaminator of waterways and William Handcock, referring to what flowed to the City Watercourse, wrote in the 1890s that "The citizens are fortunate that they are not now dependent on it, for it is so polluted by the paper-making that it has become poisonous, and cattle and horses have died from drinking it. Sometimes it is the same colour as porter".

The commissioners of the Rathmines & Rathgar and Pembroke Townships brought the first main drainage system in the Dublin area into operation in 1881. Then, to encourage development and cleanse the Dodder, they laid the first Dodder Valley sewer in the 1880s. This interceptor, still in use today, flows from Orwell Bridge to Beatty's Avenue in Ballsbridge, where it originally crossed the river to meet the main Rathmines and Pembroke trunk sewer behind Estate Cottages, off Shelbourne Road. As noted earlier, this first Dodder Valley sewer originally crossed the Dodder on piers, between which branches and other debris became entangled in times of flood. In 2000, a siphon under the Dodder replaced the pipe on the piers.

Pollution originating upstream of where a river crosses the boundary from one local authority to another has often been the cause of concern and disputes. Even after laying the first Dodder Valley sewer, the authorities in Rathmines and Pembroke were regularly and rightfully roused when the river was rendered foul by matter that originated upstream. Despite the provisions of the 1876 Pollution of Rivers Act, such incidences remained difficult to resolve until pollution control became more easily enforceable under legislation of more than a hundred years later: the Local Government (Water Pollution) Act of 1977 and its amending follow-up, enacted in 1990

In this environmentally sensitive age, dumping in rivers by opportunistic and anti-social individuals—always a serious problem—has become more widely noticed than heretofore. Hardware such as traffic cones and supermarket trolleys are relatively easy to retrieve, but litter, domestic refuse and substantial fly dumping are more difficult to remove and the results more costly and labour intensive to rectify

AMENITY

The overall amenity value of the Dodder has long been recognised. In the late nineteenth century, Handcock described its plant life and other natural features in detail. Among several other features, he described a spring close to a cottage near Friarstown called Ferndale, at that time a ruin. This spring, which had a quantity of carbonate of lime in solution, trickled through the moss and grass and, encrusting the delicate stems and leaves in a short space of time, turned the whole into a beautiful petrifaction, the upper portion being living moss, while underneath it was hard stone. Some of the specimens recovered by Handcock rivalled the finest coral. Every leaf and fibre of the delicate moss, he wrote, was transformed to durable stone. After 30 years' exposure to the weather, these were as perfect as when removed. Some of the blocks weighed several hundredweights, and were unrivalled for rockery work.

The importance of the Dodder as a prime amenity was recognised in a report prepared for Dublin Corporation in 1941. This important document, produced by a team led by the eminent planner Professor Patrick Abercrombie, proposed the creation of parkway strips along the Tolka and Dodder. "In the case of the Dodder" wrote Abercrombie, "it is a simple matter with the aid of a few light foot-bridges to give direct access for pedestrians from Ballsbridge to the green belt. In addition to these riverside walks, it is proposed to form small wedges of open land and to preserve green strips along certain major roads, the details and exact location of which would be decided at a later stage."

Also in the 1940s, architect Noel Moffett proposed a scheme for a riverside park for the Dodder, stretching ten miles from the mountains to the sea. Much later, development of a linear park along the river took place in stages over several years. A detailed study undertaken in 1985 laid down the pattern for much subsequent environmental work. The City Council's Parks Department sees the Dodder as an organising element of the recreational open space on Dublin's southern boundaries, extending from the sea to the mountains where the river rises.

In his address to the Institution of Engineers of Ireland on 16th November 1987, Jack Keyes identified the groups most concerned with the amenities of the river: fishermen, walkers, joggers, ornithologists, cyclists and nature lovers.

Handcock recorded the river as abounding with fine trout, which were sometimes illegally harvested by local people who threw quicklime into the river killing all fish in that stretch of the river. This happened when there was low flow in the river. The stock, however, was quickly replenished from the waters above Castlekelly, which were preserved. Other species included eels, sticklebacks, locheen or gudgeon, and minnows. The Dodder is still fished and, while it is a popular and much appreciated amenity, access to the Bohernabreena Reservoirs obviously has to be controlled. Visitors must observe a set of simple rules and abuse of the water, fires and careless car parking are among the potential problems.

The anglers who fish the two Bohernabreena Reservoirs are allowed only to use flies—worms are banned. The Dodder Anglers Association, a very active organisation with 1,400 members, patrol the reservoirs and the river to stop illegal fishing. They also restock the river every year. The trout fish stock in the reservoirs and the river upstream, flowing into the reservoir, is the native Irish breed. It is one of a small number of locations in Europe where the trout have not cross-bred with farmed trout.

Every effort has been made to encourage the growth of flora and fauna. Trees of several species grow near the Dodder, especially in the upper reaches. Among others, there are fir, dale, ash, oak, elm, silver birch and chestnut.

In her address to the January 2008 Dodder workshop, Dr. Mary Tubridy described the rich biodiversity of the Dodder upstream of the Ballsbridge Weir as far as Bushy Park. Of the 75 habitats identified in terrestrial locations in Ireland, twenty-two were found in the environs of this stretch of the Dodder. The semi-natural vegetation is dominated by native species including several rare species, all associated with woodlands. Seventy-one of Ireland's 120 bird species are here, including pheasants, goldfinch, kingfishers, cranes, hawks and herons. Otters are common, six species of bat have been identified and the rich spectrum of wildlife also includes deer. The river, Dr. Tubridy said, acts a corridor for wildlife and the presence of certain species is a sign of good water quality.

The City Council's Parks Department has noted that the River Dodder corridor supports a rich variety of birds, animals and insects. It sees the river as a special wildlife habitat comprising fast-flowing areas of water, slow-moving pools, ponds and flooded marshy areas around its banks. Attention is also drawn to the Dodder's former role as an industrial river, still dotted with many relic millstreams, weirs, sluices and old factories. It notes that the river connects numerous sports grounds along its length as well as affording fishing to members of the Dodder Angling Association. In managing the spaces along the river, the Parks Department's objectives include facilitating access, while safeguarding and enhancing the natural characteristics of the river for recreation and conservation.

A new and serious threat to the Dodder and its environment was identified in 2004. Signs of a major infestation of Japanese Knotweed and Himalayan Balsam along the stretch between Clonskeagh and Londonbridge were noticed. There was a danger that these highly invasive plants, which dominate all indigenous plant and insect life, would destroy the eco balance of the river. These growths have also been known to undermine walls and pipes and even break through concrete. Urgent steps were taken to deal with this menace and to watch for signs of future invasions.

The confluence of the Dodder with the Liffey at Ringsend, known as the Gut, is in an area of great complexity. Some historical details of the original village and its growth have already been outlined in the section on Dodder Villages.

The Grand Canal is entered from the Liffey through the outlet of the Dodder. In the last decade of the eighteenth century, the Grand Canal Docks were constructed at this location. The two basins, totalling ten hectares (25 acres) in extent that comprise the complex are entered through the Camden, Buckingham and Westmoreland locks. These and the names of all the adjoining quays are a reminder of those who were prominent in government and public life at a time of great prosperity over two hundred years ago.

The Grand Canal Docks were never a commercial success and by the 1830s were reported to be "nearly unused". Their most intensive use was probably by colliers serving the Works of the Alliance and Dublin Consumers Gas Company and the Dublin United Tramways Company's generating station; both of which are long since gone. But, after years of virtual abandonment, the Grand Canal Docks are now high on the list of amenities adjoining the Dodder.

BIBLIOGRAPHY

Proposed Formation of Reservoirs on the River Dodder
(Robert Mallet, 1844)

Rathmines and Rathgar Township Waterworks in Glenasmole
(Arthur Tyrell 1888)

Anna Liffey – The River of Dublin (J de Courcy)

Cleansing Rural Dublin (Frank Cullen)

Natural History of Ireland (Gerard Boate)

History of Tallaght (William Handcock)

The Rivers of Dublin (CL Sweeney)

Glenasmole Roads (Patrick Healy)

Rathfarnham Roads (Partick Healy)

Down the Dodder (Christopher Moriarty)

The Neighbourhood of Dublin (St. John Joyce)

The Liffey in Dublin (JW de Courcy)

Dublin City and County – From Prehistory to Present (Aalen
and Whelan, editors).

Society and Settlement in the valley of Glenasmole
1750–1900 (William Nolan)

Dublin's Suburban Towns (Seamus O'Maitiu)The
Dublin Historical Record

Gilbert Library

National Achieves of Ireland

National Library of Ireland

National Gallery of Ireland

Drainage Division Records, Dublin City Council

Waterworks Records, Dublin City Council

Survey and Mapping Services, Dublin City Council

River Dodder Study: Flood Management
Assessment, Dublin City Council

Donnybrook – A History (Beatrice M. Doran)

Harolds Cross (Joe Curtis)

The Wicklow Military Road (Michael Fewer)

Irish Stone Bridges – History and Heritage
(Peter O'Keefee & Tom Simington)

A History of County Dublin (Francis Elrington Ball)

An Historical Sketch of the Pembroke Township
(Francis Elrington Ball)

The Industrial Archaeology of Northern Ireland (WA McCutcheon)

Printing and Bookselling in Dublin 1670–1800 (James W Phillips)

Sources of Employment and Housing in Rathfarnham from 1850 to
1911 (Stephen Browne)

Behind the Scenes (Ernie Shepherd)

Clonskeagh – A Place in History (Martin Holland)

The Development of Milling Technology in Ireland
c.600-1875 (Colin Rynne)

A Survey of Irish Flour Milling 1801-1922 (Andy Bielenberg)

Thomas Edmondson and The Dublin Laundry (Mona Hearn)

Maps and Map-Making in Local History (Jacinta Prunty)

South Dublin County Council Library

A Guide to Early Irish Law
(Dublin Institute for Advanced Studies, 1988) (Fergus Kelly)

Early Irish Farming
(Dublin Institute for Advanced Studies, 1997) (Fergus Kelly)

AUTHORS' NOTES

In 1998 I was appointed as Senior Engineer in charge of the Design Section (water and drainage) of the Engineering Department of Dublin City Council. The upgrading of the spillways in Bohernabreena was one of the ongoing projects I was in charge of. This led to many visits to the waterworks at Bohernabreena. I was fascinated with the design of the waterworks, the crisscrossing of the water channels in the valley and the techniques used by the Victorian Engineers in separating the coloured mountain bog water from the clean water lower down in the valley. The water from the upper catchment flowed into the lower reservoir and the waters collected lower down the catchment flowed into the upper reservoir. I came across the 1844 paper *"Proposed Formation of Reservoirs on the River Dodder"* by Robert Mallet on providing a constant supply of water to the mill owners on the Dodder and Arthur Tyrrell's 1888 paper on the *"Rathmines and Rathgar Township Water Works"* in Glenasmole. It took me about four years to understand the design principals and the construction of the original Victorian Waterworks. Joe McLoughlin, who had been in charge of the waterworks for 31 years pointed out some of the earlier workings of the waterworks.

My wife's family lived in Dundrum and she was told that one of her ancestors had been a miller in Dundrum. She did not have any more details as to the location or type of mill he owned. This led me to inquire as to the nature and the various types of mills in the Dodder catchment.

I decided to write up the history of the waterworks at Bohernabreena and I persuaded Michael Corcoran to research the mills on the Dodder and Poddle. I researched the mills in the Owendoher catchment and discovered that my wife's ancestor John Garvan had owned the Willowbrook Flour Mill on Whitechurch Road, Rathfarnham from 1873 to 1882.

This book has taken over nine years to compile. Helpful suggestions from many people led to the discovery of the many books and papers referred to in this book.

I have had great encouragement and assistance from my wife Mairead and from my family in compiling this book. Melissa and Colm taught me the basic skills of *Photoshop* used to enhance the drawings, maps and photographs. Some of the more difficult images I left to them. Colm prepared a number of drawing used in the book. The sketches of the water mills were painted by Aisling.

I want to thank the many people and organisations who supplied me with photographs, drawing, maps and sketches and waived their copyright fees.

Lastly I want to thank Michael Phillips, City Engineer for his continuing encouragement to complete this book and Dublin City Council for publishing this book.

DON MCENTEE

June 2015

As a new and very junior Dublin County Council clerk in 1948 I was assigned to the Planning Department, where I was offered training as a draughtsman, in those days given on the job. This very satisfying process also brought me into close contact with many old drawings and maps, stimulating my interest in history

There was much upheaval in the planning regime during the 1950s and 1960s, culminating in joint control of the city and county planning departments. Mandatory draft planning proposals for Dublin were put on display for several months in 1967. This took place in St Mary's Church (the Black Church), off Dorset Street, at that time a municipal exhibition centre. I was one of two draughtsmen present each day to answer questions and interpret the drawings for visitors. My polished senior colleague became known as the Parish Priest, with me as his curate.

I next went to a more senior Dublin Corporation post in the Valuer's Office, which bought and sold property on behalf of the city. At that time, great land banks were being assembled to provide sites for housing, much of the property being in the Tallaght area. This was the first educational contact with the Dodder for a Northsider far more familiar with the Tolka. There was much survey work, checking boundaries and preparation of maps, much of which involved ancient maps, a further source of learning.

During that period, many 199-year leases in the City Estate portfolio, originally granted around 1770, came up for renewal. Site maps were checked and compared, premises inspected and sometimes surveyed. Some buildings were bought, necessitating the production of floor plans. Much of this work dissipated in 1969 when the work of the Valuer was partly concentrated in the Main Drawing Office (now Survey and Mapping), where I returned to Imperial measurements from the Metric System used by the Valuer since 1967.

Next came a transfer, in 1971, to the Drainage Division, where I was one of a team whose task was to update the sewer records on the Ordnance Survey maps. Due to a severe staff shortage over many years, these important maps needed urgent revision – and such was the shortage of Ordnance sheets to the new scale of 1:1000 that we often made our own, which sufficed until the new issues became available. There were many very old drawings in this department which were of great historical interest.

Following retirement in 1995, I wrote a history of the Dublin tramways and was exhilarated shortly afterwards on being invited back, part-time, to the Drainage Division to record and catalogue thousands of old drawings on to the computer. I also wrote a history of water and drainage in Dublin and tackled several other tasks. It was a great honour to be invited by Don to join him on the Dodder project, and I am sorry that an unexpected two-year health hiatus delayed his work. But it was great to get re-involved!

MICHAEL CORCORAN

June 2015

LIST OF ILLUSTRATIONS

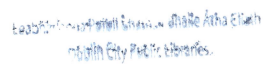

ꟽNDEX